Urban Water Resources

Monzur A. Imteaz
Swinburne University of Technology
Hawthorn, Victoria, Australia

CRC Press
Taylor & Francis Group
Boca Raton London New York

CRC Press is an imprint of the
Taylor & Francis Group, an **informa** business

A SCIENCE PUBLISHERS BOOK

CRC Press
Taylor & Francis Group
6000 Broken Sound Parkway NW, Suite 300
Boca Raton, FL 33487-2742

First issued in paperback 2020

© 2019 by Taylor & Francis Group, LLC
CRC Press is an imprint of Taylor & Francis Group, an Informa business

No claim to original U.S. Government works

ISBN-13: 978-1-138-33989-7 (hbk)
ISBN-13: 978-0-367-77927-6 (pbk)

Visit the Taylor & Francis Web site at
http://www.taylorandfrancis.com

and the CRC Press Web site at
http://www.crcpress.com

Dedicated to
My wife

Acknowledgements

The author would like to express infinite thanks to God, the creator of the universe, through whose mercy this book is written to its current form. Also, the author expresses gratefulness to the people and publisher who have provided support during the preparation and publication of the book. Special gratitude to Humaira Imteaz, Miqdaad Imteaz, Yunus Ahmed and Jareer Imteaz for their help and thorough review of the book in regard to language and drawing graphics.

Preface

Adverse impacts of climate change are evident in many regions of the world and are likely to worsen in the future. Ever increasing urbanisation is also impacting both the quantity and quality of urban water resources. These urban water resources and components of the water cycle are likely to be affected severely. To minimise the consequences on world water resources, the development of sustainable water resources management strategies is inevitable. An integrated urban water resources management strategy is the key to maintaining sustainable water resources. A preliminary understanding of physio-chemical processes and analysis methodologies involved in each and every component of the urban water cycle is necessary. In the past, these components have been investigated and published individually. With the purpose of aiding the development of integrated urban water resources management strategies, this book endeavours to present and explain the major urban water cycle components from a single holistic platform. The book presents the introduction, analysis and design methods of a wide range of urban water components, i.e., rainfall, flood, drainage, water supply and waste water. The book starts with components and classifications of world water resources, then basic and detailed components of the hydrologic cycle, rainfall patterns and measurements, rainfall losses (i.e. evaporation, transpiration and infiltration), derivations of design rainfalls, streamflow measurements, flood frequency analysis and probabilistic flood estimations, deterministic flood estimations, unit hydrograph, flood modelling for heterogeneous catchments, principles of open channel hydraulics, critical flow and flow classification indices, open channel flow profiles, uniform flow in open channel and open channel design, hydraulic modelling, estimation of future population and domestic water demand, design of water supply systems, water treatments, wastewater quantification, wastewater treatments, stormwater drainage, water conservation and recycling, and sustainable urban design.

Monzur A. Imteaz

Contents

About the Author

Dr. Monzur Imteaz is an Associate Professor in the Department of Civil & Construction Engineering at Swinburne University of Technology, Melbourne. He obtained his Bachelor of Civil Engineering degree from the Bangladesh University of Engineering & Technology and then his Master of Engineering degree from the Asian Institute of Technology, Thailand. Through his excellent academic performance, he was able to secure a Japanese government scholarship to do his PhD in Japan. He obtained his Ph.D. in 1997 from Saitama University, Japan. Upon completion of his PhD, he started working with the Institute of Water Modeling (Bangladesh), where he worked in collaboration with the Danish Hydraulic Institute (DHI) and he spent some time at DHI headquarter in Denmark. Later, he completed his post-doctoral research at University of Queensland, Brisbane, Australia. Before joining at Swinburne, he was involved with several Australian local and state government authorities in Queensland, New South Wales and Victoria. At Swinburne, Dr Imteaz is teaching subjects 'Urban Water Resources' and 'Integrated Water Design'. Also, he has been actively involved with various researches on sustainability, water recycling, modeling pollutants transport and treatment, developing decision support tools, and rainfall forecasting using Artificial Neural Networks.

To date, Monzur is the author of 1 edited book, 11 book chapters, 117 journal papers and 95 conference papers. Due to his excellent research outcomes in the field of 'sustainability', in 2008 he was awarded with Swinburne's 'Vice-Chancellor's Sustainability Award'. Again, in 2015, in recognition to his excellent teaching, Monzur was awarded Swinburne's 'Vice Chancellor's Teaching Excellence Award'. To date, he has successfully supervised eleven PhD students to completions. In recognition of his service to the international research community, Monzur was appointed as an Editorial Board member for the renowned journals, 'Resources, Conservation and Recycling' and 'International Journal of Hydroinformatics'.

Introduction

1.1 World Water Resources

Water is a renewable resource and it exists in different forms. The total quantity (including all the forms) of water in the world is approximately 1.36 billion cubic kilometres (km^3). However, out of this vast amount of water almost 96.5% is contained in the oceans, which is saline and not readily consumable. Out of the remaining 3.5%, almost 50% is contained as ice sheets/caps and permafrost in the Polar Regions and on mountain tops. Another 48% of the non-oceanic water is stored as groundwater, either as extractable groundwater or as soil moisture. Only about 0.013% of total water is available in lakes, rivers and reservoirs. Out of the total share of groundwater (1.69%), again 55% is saline and not readily consumable. To visualise these proportions in an easy way, if we consider the world's total water volume equal to a 15 liters water bottle (usually used in drinking water fountains), a 375 mL can of cold drinks would be the volume of total fresh water and a teaspoon of 5.4 mL capacity would be able to hold the fresh water available in lakes, rivers and reservoirs. Table 1.1 shows the detailed distribution of world water resources in different forms.

Table 1.1. Distribution of world water resources

Source of water	Volume of water ($\times 1000\ km^3$)	% Total water
Oceans	1,338,000	96.5
Ice and Snow	24,364	1.76
Groundwater	23,400	1.69
Lakes	176.4	0.013
Atmosphere	12.9	0.00093
Swamps	11.47	0.00083
Rivers	2.12	0.00015

1.2 Classifications of Water Resources

In general, water resources can be defined as the sources of water which are potentially useful and available for fulfilling basic needs. These can be classified based on their sources as follows:

(a) Surface water
(b) Ground water
(c) Spring water
(d) Frozen water
(e) Ocean water

These can be further classified in to two categories: i) readily available and ii) non-readily available. Among the five mentioned water resources, only 'surface water' and 'spring water' are readily available. 'Ground water' requires some extraction processes, some 'frozen water' from mountain tops melts during the summer and flows through the rivers and 'ocean water', though to make it readily available requires an expensive treatment process (desalination).

1.3 Climate and Climate Change

The climate of a particular area is the usual pattern of weather in that specific area over long periods of time, preferably 30 years or more. Although several factors/variables, such as temperature, humidity, atmospheric pressure, wind, and precipitation, contribute to the assessment of climate in a particular area/region, temperature and precipitations are the most familiar features of the climate of a particular area/region. Of these two most familiar features, precipitation directly influences many water/civil engineering activities and decision making. Temperature has direct influence on building thermal comfort and indirect influence on different water engineering activities such as water demand and water loss calculations in the irrigation sector, which requires the consideration of evaporation and depends on temperature. A region's climate is the ultimate outcome of interactions among five major climate systems, which are: Atmosphere, hydrosphere, cryosphere, lithosphere, and biosphere. In the following chapter, the interactions three of these systems (atmosphere, hydrosphere and lithosphere) will be discussed in detail.

Water is a salient component of the climate system and plays a significant role in climate variations. With ever-increasing anthropogenic activities and emissions of greenhouse gases, scientists have been predicting global warming and disruptions of traditional flow patterns of the various components (rainfall, evaporation, snow-melting, etc.) of the climate system. There are succinct pieces of evidence in different parts of the world, which reveal that adverse impacts of climate change and global warming are already occurring. Climate change will have significant impacts on water resources around the world because of the close connections between the climate and natural water cycle. Rising temperatures will increase evaporation from water bodies and lead to increases in rainfall. However, scientists predict that this increase in rainfall will not be evenly distributed both temporally and spatially. There will be significant regional and temporal variations in rainfalls. In some parts of the world there will be heavy rainfall, whereas some other parts of the world will suffer from a lack of rainfall, which will cause both droughts and floods to occur more frequently. In mountainous areas, dramatic changes in snowfall and snow melt are expected, causing a shift of the peak time and magnitude of river flows. In general, the net amount of snow on mountain tops will reduce, which will cause shortages of water supply for the nations dependent on river flows generated from upstream mountains. Higher temperatures will cause thermal expansion of the ocean water-body, causing the sea-level to rise. Sea-level rise will have devastating impacts on the nations living in many coastal areas. Sea-level rise also means the increase in sea surface, which will contribute to higher evaporation in addition to elevated evaporation induced by higher temperature. Climate change will also increase the demand for farm irrigation and water consumption. In summary, climate change will have significant impacts on the water sector through the natural water cycle, water availability, water demand, and water allocation at the global, regional, and local levels.

1.4 Seasonality Index

Engineering designs are likely to vary from region to region, more specifically from climate to climate. To provide a suitable design for a certain region/climate, engineers often need to define the climate and its sub-systems. Seasonality Index (SI) is one of such indicators expressing variations of rainfalls. SI is the average monthly rainfall variations within a year for a particular locality, calculated through the following equation as proposed by Summer (1988).

$$\text{SI} = \frac{1}{R} \sum_{j=1}^{12} \left| X_j - \frac{R}{12} \right| \tag{1.1}$$

where 'X_j' is the mean rainfall of a month 'j' and 'R' is the mean annual rainfall for a particular location/city. Basically, a high SI does not mean high amounts of rainfall, rather it means higher variations of rainfalls within the months. Even a small amount of total annual rainfall may exhibit very high variation within the months, leading to a high SI value.

Hypothetically, the SI can vary from 0-1.83; an equal amount of rainfall for all the months within a year will lead to a SI value of '0', whereas all the rainfall occurring in only one month of a year will lead to a SI value of '1.83'. Aryal et al. (2009) provided a classification of rainfall regimes based on SI values, as shown in Table 1.2.

Table 1.2. Seasonality Index classes and related different rainfall regimes

Rainfall regime	Seasonality Index
Very equable	≤ 0.19
Equable but with a definite wetter season	0.20–0.39
Rather seasonal with a short drier season	0.40–0.59
Seasonal	0.60–0.79
Markedly seasonal with a long drier season	0.80–0.99
Most rain in 3 months or less	1.00–1.19
Extreme, almost all rains in 1-2 months	≥ 1.20

1.5 Drought Index

Weather does fluctuate, even though a general average pattern exists, however, at times, variation from year to year is quite high. In regard to water/rainfall, some years have higher than average rainfall, whereas some other years have lower than average rainfall. If this variation is very high, it causes enormous problems; i.e. very high rainfalls cause floods and very low rainfalls cause drought. Water managers and stakeholders need to measure/assess the extent of wetness/drought. There are several drought indices used by different authorities/individuals. The most widely-used drought index is Standardized Precipitation Index (SPI), which characterizes weather condition based on real rainfall data. Basically, SPI calculates deviation of observed precipitation for a certain period (month, season or year) from a long-term mean for the same period normalized by the standard deviation of long-term precipitation values for the same period. In simple form, the SPI can be defined as follows:

$$\text{SPI} = \frac{(p - p^*)}{\sigma_p} \tag{1.2}$$

where 'P' is precipitation at a certain time period, 'P^*' is the mean precipitation for the same time period from a long-term precipitation values and 'σ_p' is the standard deviation of the long-term precipitation values for the same period. Although it is generally known as 'drought index', it expresses all the 'dry or extremely dry conditions', 'average conditions' and 'wet or extremely wet conditions'. Negative SPI values correspond to a dry to extreme dry (values from −1 to −3) period, positive SPI values correspond to a wet to extremely wet (values from 1 to 3) period, and values between −0.99 to 0.99 correspond to a normal or near normal condition. More details with examples can be found on the National Center for Atmospheric Research (NCAR) website (UCAR, 2018). Table 1.3 shows the classification of SPI values.

Table 1.3. Classification of SPI values

SPI value	Weather class
≤ −2.0	Extremely dry
−1.99 to −1.50	Severely dry
−1.49 to −1.0	Moderately dry
−0.99 to 0.99	Near normal
1.0 to 1.49	Moderately wet
1.5 to 1.99	Very wet
≥ 2.0	Extremely wet

Worked Example 1

Monthly average rainfalls (mm) for a city are given below:

20, 22, 25, 35, 54, 25, 65, 50, 45, 40, 30, 25

Calculate seasonality index for the rainfall pattern for this city.

Solution:

From the given rainfall data, annual average rainfall (R) for the city can be calculated by adding all the monthly average values,

R = 20+22+25+35+54+25+65+50+45+40+30+25 = 436 mm

Now, calculate the absolute values of ($X_j − R/12$) for each month:

$$|X_1 − R/12| = |20 − 436/12| = 16.33$$
$$|X_2 − R/12| = |22 − 436/12| = 14.33$$
$$|X_3 − R/12| = |25 − 436/12| = 11.33$$
$$|X_4 − R/12| = |35 − 436/12| = 1.33$$
$$|X_5 − R/12| = |54 − 436/12| = 17.67$$
$$|X_6 − R/12| = |25 − 436/12| = 11.33$$
$$|X_7 − R/12| = |65 − 436/12| = 28.67$$
$$|X_8 − R/12| = |50 − 436/12| = 13.67$$
$$|X_9 − R/12| = |45 − 436/12| = 8.67$$
$$|X_{10} − R/12| = |40 − 436/12| = 3.67$$
$$|X_{11} − R/12| = |30 − 436/12| = 6.33$$

$$| X_{12} - R/12 | = | 25 - 436/12 | = 11.33$$

Now, taking sum of all the above values,

$$\sum_{j=1}^{12} |(X_j - R/12)| = 144.67$$

Now, Seasonality Index for the city = $144.67/R = 144.67/436 = 0.33$

Worked Example 2

For a particular city, monthly rainfall values (mm) are available for the whole year except for one month (Month '7' value is missing). Also, total annual rainfall amount is missing. However, SI value for that particular year is known as 0.30. From the given data, calculate the missing monthly rainfall value. Monthly rainfall values for the city are (X represents the missing value):

20, 25, 30, 35, 40, 50, X, 50, 40, 30, 25, 20

Solution:

As we are unable to calculate the total 'R' value, due to the missing value (X), we can't get a direct solution. Rather, we have to use trials to find out the solution. We have to assume the missing value (X) and then calculate the SI value with the assumed monthly value. We have to repeat this process until we get a SI value close to "0.30".

Let's assume the missing value, $X = 70$. Now, calculate the annual rainfall value, $R = 20+25+30+35+40+50+70+50+40+30+25+20 = 435$ mm

Now, calculate the absolute values of $(X_j - R/12)$ for each month:

$$| X_1 - R/12 | = | 20 - 435/12 | = 16.25$$
$$| X_2 - R/12 | = | 25 - 435/12 | = 11.25$$
$$| X_3 - R/12 | = | 30 - 435/12 | = 6.25$$
$$| X_4 - R/12 | = | 35 - 435/12 | = 1.25$$
$$| X_5 - R/12 | = | 40 - 435/12 | = 3.75$$
$$| X_6 - R/12 | = | 50 - 435/12 | = 13.75$$
$$| X_7 - R/12 | = | 70 - 435/12 | = 33.75$$
$$| X_8 - R/12 | = | 50 - 435/12 | = 13.75$$
$$| X_9 - R/12 | = | 40 - 435/12 | = 3.75$$
$$| X_{10} - R/12 | = | 30 - 435/12 | = 6.25$$
$$| X_{11} - R/12 | = | 25 - 435/12 | = 11.25$$
$$| X_{12} - R/12 | = | 20 - 435/12 | = 16.25$$

Now, taking sum of all the above values,

$$\sum_{j=1}^{12} |(X_j - R/12)| = 137.5$$

Now, Seasonality Index for the city = $137.5/R = 137.5/435 = 0.32$

As the calculated 'Seasonality Index' turned out to be higher than the given 'Seasonality Index', 0.30. We need to proceed with further trials. The calculated index is higher than the given index, which means that we have assumed a higher missing value (X) than its real value. As such, the next assumption for 'X' should be lower than the earlier assumed value. Let's assume $X = 60$. Now, follow the same procedure:

$R = 20+25+30+35+40+50+60+50+40+30+25+20 = 425$ mm

Now, calculate the absolute values of ($X_j - R/12$) for each month:

$$|X_1 - R/12| = |20 - 425/12| = 15.42$$
$$|X_2 - R/12| = |25 - 425/12| = 10.42$$
$$|X_3 - R/12| = |30 - 425/12| = 5.42$$
$$|X_4 - R/12| = |35 - 425/12| = 0.42$$
$$|X_5 - R/12| = |40 - 425/12| = 4.58$$
$$|X_6 - R/12| = |50 - 425/12| = 14.58$$
$$|X_7 - R/12| = |60 - 425/12| = 24.58$$
$$|X_8 - R/12| = |50 - 425/12| = 14.58$$
$$|X_9 - R/12| = |40 - 425/12| = 4.58$$
$$|X_{10} - R/12| = |30 - 425/12| = 5.42$$
$$|X_{11} - R/12| = |25 - 425/12| = 10.42$$
$$|X_{12} - R/12| = |20 - 425/12| = 15.42$$

Now, taking sum of all the above values,

$$\sum_{j=1}^{12} |(X_j - R/12)| = 125.83$$

Now, Seasonality Index for the city = $125.83/R = 125.83/425 = 0.30$

As this value (0.30) is matching with the given SI value, the missing month's rainfall value (X) would be '60' mm.

As absolute values are taken for ($X_j - R/12$) values, a value of '13' for the 'X' will also produce same SI value, i.e. 0.30. As such, there are two possible values of 'X' for such scenario. In this scenario, we have to apply our judgement in order to choose the correct missing value and, ideally, we should take a value which is closer to the long term average rainfall for that particular month for the same city.

Note: As this process is a trial/repetitive process, for such analysis it is recommended that we use a spreadsheet in order to achieve a quick solution.

Worked Example 3

For a particular city, annual rainfall data for sixteen years are given below. In the most recent year, an annual rainfall of 325 mm was observed. Calculate the SPI value for the most recent year and categorise the weather of that year. Sixteen years annual rainfall values (in mm) are:

400, 350, 450, 510, 520, 430, 520, 360, 370, 450, 470, 460, 550, 600, 630, 650

Solution:

First of all we need to calculate, mean precipitation (P^*), which is:

$$= (400+350+450+510+520+430+520+360+370$$

$$+450+470+460+550+600+630+ 650)/16 = 482.5 \text{ mm}.$$

Now, we need to calculate the standard deviation (σ_p) for the given data set. The equation to calculate standard deviation is:

$$S = \sqrt{\frac{\Sigma(X_i - M)^2}{N}}$$

where 'X_i' are the individual data from the series, 'M' is the mean and 'N' is the total number of data.

For this particular case, N = 16 and M = 482.5 (calculated). To be able to calculate 'S', we need to calculate sixteen $(X_i - M)^2$ values, as shown below:

$(X_1 - M)^2 = (400 - 482.5)^2 = 6806.25$
$(X_2 - M)^2 = (350 - 482.5)^2 = 17556.25$
$(X_3 - M)^2 = (450 - 482.5)^2 = 1056.25$
$(X_4 - M)^2 = (510 - 482.5)^2 = 756.25$
$(X_5 - M)^2 = (520 - 482.5)^2 = 1406.25$
$(X_6 - M)^2 = (530 - 482.5)^2 = 2756.25$
$(X_7 - M)^2 = (520 - 482.5)^2 = 1406.25$
$(X_8 - M)^2 = (360 - 482.5)^2 = 15006.25$
$(X_9 - M)^2 = (370 - 482.5)^2 = 12656.25$
$(X_{10} - M)^2 = (450 - 482.5)^2 = 1056.25$
$(X_{11} - M)^2 = (470 - 482.5)^2 = 156.25$
$(X_{12} - M)^2 = (460 - 482.5)^2 = 506.25$
$(X_{13} - M)^2 = (550 - 482.5)^2 = 4556.25$
$(X_{14} - M)^2 = (600 - 482.5)^2 = 13806.25$
$(X_{15} - M)^2 = (630 - 482.5)^2 = 21756.25$
$(X_{16} - M)^2 = (650 - 482.5)^2 = 28056.25$

Now, sum of all the above values, $\Sigma(X_i - M) = 129300$

So, $\sigma_p = S = \sqrt{(129300/16)} = 89.9$

Now, using Equation 1.2, SPI $= (P - P^*)/\sigma_p = (325 - 482.5)/89.9 = -1.75$

As per Table 1.3, the most recent year was a 'severely dry' year.

Note: These lengthy calculations can be simplified using 'STDEV.P' in Excel spreadsheet.

References

Aryal, S., Bryson, B., Campbell, E. and Li, Y. (2009). Characterising and Modelling Temporal and Spatial Trends in Rainfall Extremes. Hydrometeorology 20(1): 241–253.

Summer, G. (1998). Precipitation: Process and Analysis. John Wiley and Son, Chichester.

UCAR (2018). University Corporation for Atmospheric Research Climate Data Guide. https://climatedataguide.ucar.edu/climate-data/standardized precipitation-index-spi (accessed on 10 May 2018).

Hydrologic Cycle and
Rainfall-Runoff Processes

2.1 Hydrological Cycle and Systems

Water can stay within the earth system in three different forms, i.e. liquid, solid and gaseous. In reality, the total amount of water (combining all the forms) within the earth system remains the same over time. In general, water remains in one of the systems; i) Atmospheric system ii) Lithospheric system and iii) Oceanographic system. However, the amount of each individual component changes as water continuously changes from one state to another. In general, the continuous occurrence and movement of water within these systems can be defined as 'hydrologic cycle' or 'water cycle'. More precisely, the occurrence, distribution and movement of water in the natural environment is a cyclical process, known as the 'hydrologic cycle'.

   Figure 2.1 shows the schematic diagram of interactions among hydrologic systems. From the atmospheric system, water moves to the lithospheric system through precipitation, and from the lithospheric system some water goes back to the atmospheric system through evaporation and transpiration. From the lithospheric system, water moves to the oceanographic system through streamflow and ground water movement. From the oceanographic system, water moves to the atmospheric system through evaporation. Again from the atmospheric system, a huge amount of water comes back to the oceanographic system through precipitation. Among all the above-mentioned systems, hydrologists and civil engineers are mainly concerned with the lithospheric system. Figure 2.2 shows a schematic diagram of the 'hydrologic cycle', where all the major components, i.e. precipitation, evaporation, transpiration, surface runoff, infiltration and groundwater movement, are shown. These components are described in the following sections. Figure 2.3 shows the detailed

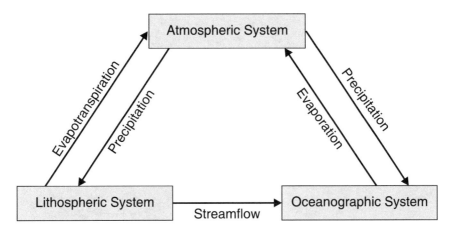

Figure 2.1. Schematic diagram of interactions among hydrologic systems

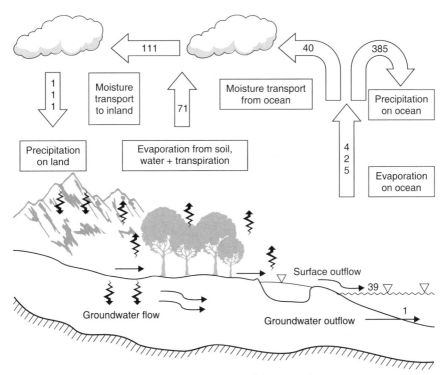

Figure 2.2. Schematic diagram of the hydrologic cycle
(Numbers are in thousands km^3 per year)

schematic diagram and components of the hydrologic cycle with their flow directions (Chow et al., 2017).

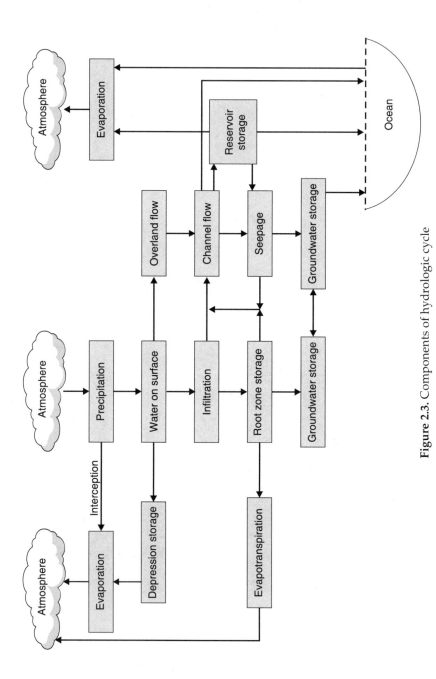

Figure 2.3. Components of hydrologic cycle

2.2 Water Balance

In general, water balance is a basic principle, which can be explained in a simplistic way: "Consider a bucket having an outflow/outlet provision with a certain capacity and water is being poured into the bucket, allowing the outlet to drain the water from the bucket. Now, in a certain time period, the difference between the inflow and the outflow would be equal to the change of storage volume in the bucket." By applying this basic principle, several issues of water resources and hydrology can be solved easily. Often, the ocean and/or groundwater storage is considered as the bucket defined in the above simplistic definition. Considering that the amount of water in the ocean and underground remain the same within a long-term period, the global average annual water movements are shown in Figure 2.2. In this figure, the numbers are shown as water movements in thousands km^3 per year. Among the annual transports of water, the highest amount is 425,000 km^3 as evaporation from ocean surface, however, out of this amount, approximately 91% (385,000 km^3) comes back to the ocean again as precipitation. Only about 9% (40,000 km^3) of the huge ocean evaporation travels inland, which joins with the evapotranspiration from the land and trees on the earth, and ultimately causes an amount of 111,000 km^3 falling on the earth's surface as precipitation. Out of this precipitation amount on land, 64% (71,000 km^3) goes back to the atmosphere as evapotranspiration (evaporation from land and water, plus transpiration from plants). Out of the total precipitation amount on land, 35% (39,000 km^3) flows as surface flow and eventually joins with the ocean. Less than 1% (approximately 1,000 km^3) flows through groundwater and finally joins with the ocean. Some key features from the mentioned circulation are: Ocean surface receives 3.5 times more precipitation than the land, evaporation from the ocean surface is 3.8 times that of the precipitation on the land, and the amount of ocean evaporation transporting to the land is 36% of the precipitation on the land and 9% of the total evaporation from the ocean. In recent times, some scientists proposed to artificially alter/divert some of these natural flows as per our need. Nonetheless, whenever mankind tried to alter the natural system, it caused several unintended consequences. Table 2.1 shows the brief comparison of the world's water balance with Australia's water balance.

Table 2.1. Comparison of world water balance with Australian water balance

	World average	*Australia average*
Precipitation on land (mm/yr)	745	736
Precipitation volume on land (km^3/yr)	111,000	5,661
Evapotranspiration (% of precipitation)	64	69.3
Surface water flow (% of precipitation)	35	23.4
Groundwater flow (% of precipitation)	1	7.3

2.3 Precipitation

Water which falls from the atmosphere to the earth's surface, either in the form of liquid, particles or frozen, is defined as precipitation. Precipitation occurs when a portion of the atmosphere becomes saturated with water vapour and the atmospheric temperature drops near saturation point. At this stage, condensation occurs and 'water vapour' turns into liquid/frozen water. However, some of this condensed water may not reach the earth's surface. This is not to be considered as precipitation, since precipitation is only that portion which falls on the earth's surface. Most of the water in fog and mist do not reach to the earth surface, rather they evaporate back to the atmosphere before falling on the earth surface. As such, most of the portions in fog and mist are not considered as precipitation.

Due to condensation of saturated air, tiny water droplets are formed. Often, small particles such as salt, clay, dust, etc., help to form droplets around these particles. Tiny liquid droplets start getting bigger through collision and coalescence within droplets. Bigger droplets start falling due to gravity, however they are often resisted by rising airflow. Precipitation occurs when the falling velocity becomes larger than the rising airflow rate. Some of these falling droplets will evaporate again due to higher temperature and less saturated air underneath (near the earth's surface). Precipitation can be divided into three broad categories, based on whether it falls as liquid, liquid that freezes (upon contact with the surface) and frozen. In general, different types of precipitations are: Raindrops, ice pellets, hail, snowflakes, diamond dust and fog/mist landed on earth. Precipitation is usually measured as accumulated depth of water over a certain period of time.

With respect to mechanisms of precipitation formation, they can be categorised into three types: Convective, orographic and cyclonic precipitation/rainfall. Brief description of formation of different types of precipitation is as follows:

(a) **Convective:** This occurs when solar-heated air rises up due to lesser density; while going up, it carries water vapour with it. As it rises up, it gets cooler and eventually condensation occurs. Through condensation around particles and coalescence, raindrops become bigger and start falling as light showers to intense thunderstorms. This type of precipitation is very common in tropical areas. Figure 2.4 shows the schematic diagram of convective rain formation.

(b) **Orographic:** This occurs when wind-driven moist air is lifted over mountain (or rising terrain) barriers and moist air gets cooler and, eventually, condensation occurs. Figure 2.5 shows the schematic diagram of orographic rain formation.

(c) **Cyclonic:** This occurs due to strong circulating air flow. Due to unequal warming of earth surfaces, regions of high pressure and low pressure develop. Air starts moving from high pressure to low pressure. Circulation is occurred due to the Coriolis effect (i.e. earth's rotation). Circulating airflow extracts the moisture-laden air and eventually falls as heavy downpour.

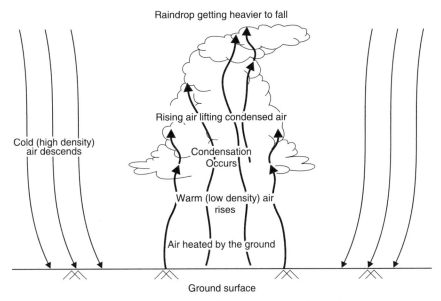

Figure 2.4. Mechanism of convective rain formation

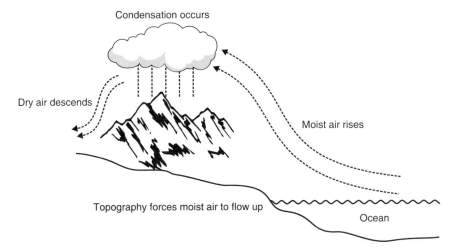

Figure 2.5. Mechanism of orographic rain formation

2.4 Precipitation Measurement

As for many water engineering analyses, historical precipitation records are imperative; measurement of precipitation has been a fundamental task for many years. As the precipitation measurement is very easy and simple, in some areas there are even precipitation records for the last hundred years. Precipitation is measured as the depth of water, which is accumulated in a bucket having vertical walls/sides left open (under sky) during rain event(s) for a specified period of time. The traditional practice was that a person visits the rain-gauge (the bucket) site at specified intervals and reads the accumulated depth of water through a ruler attached to the bucket. This manual method of measurement has been widely used for many years, especially where manpower is cheap or people are available to do this voluntarily. Figure 2.6 shows the simplest form of a rain gauge, although in real life it is customised for ease of use/reading (i.e. Figure 2.7).

With the advent of technological developments, several other sophisticated automated methods for the measurement of precipitation have evolved. As the manual reading is troublesome, especially for some remote sites (or during inclement weather), the feasible time period for such measurement has been a day (24 hours). However, with the advent of automated methods, some devices can even measure it every minute. Among the automated devices, the one frequently used is the tipping bucket type rain gauge. This device contains a small bucket, during a rain event it accumulates water and tilts when it gets full (Figure 2.8). A recorder

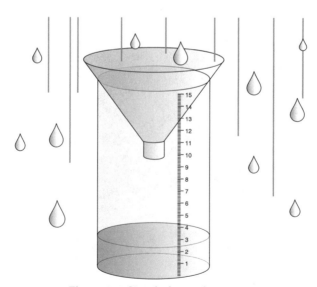

Figure 2.6. Simple form of rain gauge

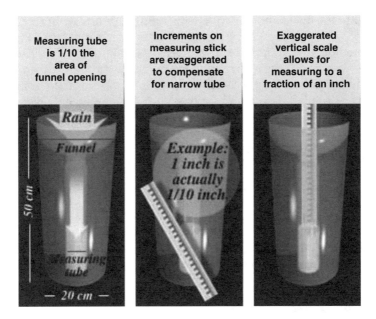

Figure 2.7. Operation and reading from simple rain gauge
(*Source*: usatoday30.usatoday.com/weather/wrngauge.htm)

Figure 2.8. Operation of tipping bucket type rain gauge
(*Source*: usatoday30.usatoday.com/weather/wrngauge.htm)

connected to the bucket holder counts the number of tilts within a certain time period (i.e. 5 minutes). Each tilt accounts for a certain depth of rainfall. By this method, precipitation can be measured automatically. Another automated method is using radar measurements through transmitting electromagnetic waves and recording reflected waves. In Australia, there are historical rainfall data for more than 19,000 stations, however, only approximately 8000 stations are currently active. Australian Bureau of Meteorology (BoM) is responsible for collecting and maintaining these data. Out of those rain gauges, BoM has got approximately 700 automated gauges. These data can be obtained from the online source: www.bom. gov.au/climate/data/index.shtml.

2.5 Rainfall Variability

Rainfall has both temporal and spatial variability. In many places these variabilities are very significant and have enormous impact on water resources planning and use.

2.5.1 Temporal Variability

In regard to temporal variability, it can be either inter-annual variability or variability within a year. Inter-annual variability can be easily assessed through statistical analysis of annual rainfall records for many years. Traditionally, from a sample of many years' records, a 10 percentile value is defined as a dry year; a 50 percentile value is defined as an average year and a 90 percentile value is defined as a wet year. Earlier, in chapter 1, a variability parameter named 'Seasonality Index' was introduced. Basically, 'seasonality index' reveals the distribution of rainfalls within a year. For the purpose of agriculture, in addition to having a significant annual average rainfall it is also important to have the rainfall uniformly (or near uniformly) distributed over the year. Otherwise, high annual rainfall concentrated within one or two month(s) is likely to cause floods. Table 2.2 shows the 'seasonality index' for different Australian cities along with their mean annual rainfalls.

2.5.2 Spatial Variability

Usually, regions near oceans facing the direction of the ocean current receive a good amount of rain; on the other hand, regions far from the ocean receive less/no rain. Spatial variations of rainfalls are well-established based on historical long-term rainfall records in different locations. Humans have selected their habitats mainly based on this free supply of water, and cities were built based on sufficient supplies of water. Figure 2.9 shows the contour of mean annual rainfalls for the whole of Australia. From the figure, it is clear that some coastal areas are not receiving much rainfall, this is because these areas are not facing any significant current direction.

Table 2.2. Seasonality index and mean annual rainfalls of different cities

City	Seasonality index	Mean annual rainfall (mm)
Brisbane	0.363	1179
Ballarat	0.260	652
Cairns	0.765	1963
Darwin	0.911	1589
Melbourne	0.170	618
Perth	0.747	760
Sydney	0.298	1266

As the rainfall measurement is very easy, hundreds of different cities have been measuring rainfalls for many years. Based on these measurements, a good estimation of long-term annual average rainfalls for different cities is established. For Australia, daily rainfalls are typically measured at 9 am, which represents daily rainfall of the previous day (measured from 9 am of the previous day).

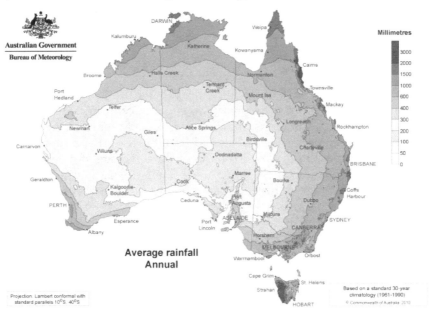

Figure 2.9. Contours of Australian average annual rainfalls
Source: www.eldoradocountyweather.com/forecast/australia/
australia-yearly-rainfall.html

Color version at the end of the book

2.5.3 Rainfall Hyetograph

Measured rainfalls are typically shown as series of bars known as 'rainfall hyetograph', where rainfall depths (typically 'mm' in Australia and 'Inch' in USA) are shown as the height of the bars. Each bar shows the rainfall amount during a selected time period (5 minutes, 10 minutes, 15 minutes ... 1 hr, 2 hr ... 24 hr, ... 72 hr). Figure 2.10 shows a typical rainfall hyetograph.

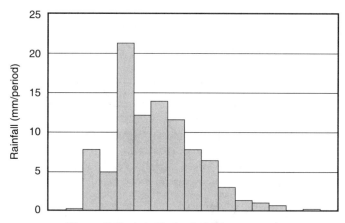

Figure 2.10. A typical rainfall hyetograph

2.5.4 Average Rainfalls

As rainfalls vary both temporally and spatially, a suitable averaging technique is required for the representation of a specific rainfall event. In regard to temporal variations, the rainfall is usually represented by the mean rainfall of a particular time period (month, year). For example, the mean annual rainfall for a particular area is the average of many years' annual rainfall values for the same area. Similarly, the mean monthly rainfall is the average of many years' monthly (for a particular month) rainfalls for the same area.

Spatial averaging is required when an average rainfall amount/intensity needs to be reported for a comparatively larger area, having more than one rain gauges/measurements within the area. Several methods are being applied for such averaging, which are described below:

(a) **Arithmetic Mean Method:** This is the simplest method, which is just taking an average of all the rain gauge measurements within the area. In some typical cases this method many not provide accurate representation of the average rainfall.

(b) **Thiessen Mean Method:** This method considers contribution of the proportionate area associated with each rain gauge station. Basically, this method is similar to the weighted average method, where the mean rainfall is calculated as per the following equation:

$$\overline{R} = \frac{A_1 R_1 + A_2 R_2 + \ldots + A_N R_N}{A_1 + A_2 + \ldots + A_N} \tag{2.1}$$

where A_N is the area where rainfall measurement is R_N (N varies from 1 to any number). For the application of this method, associated areas are delineated in a special way: The locations of all the rain gauge stations are shown as dots, then these dots are connected with each other through straight lines (Figure 2.11). Perpendicular bisectors of all these straight lines are drawn and extended until they are connected with another perpendicular bisector or the catchment/locality boundary. Now the area represented by each rain gauge station is the area surrounded by the perpendicular bisectors and the catchment/locality boundary.

(c) **Isohyetal Method:** This method first draws the rainfall contours in an area based on the point/station measurements of few rain gauge stations (Figure 2.12). Then the area between two contours

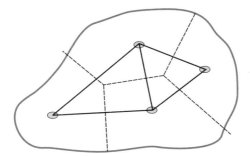

Figure 2.11. Area delineation in Thiessen mean method

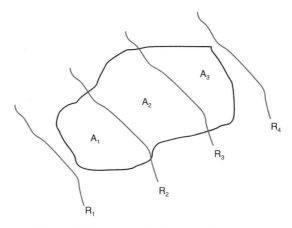

Figure 2.12. Schematic of isohyetal method

is calculated (for the outer ends if one contour is missing, then the associated rainfall for that area is the nearest contour value). Finally, the mean rainfall is calculated using the following equation:

$$\overline{R} = \frac{A_1(R_1 + R_2)/2 + A_2(R_2 + R_3)/2 + A_3(R_3 + R_4)/2}{A_1 + A_2 + A_3} \qquad (2.2)$$

2.6 Evaporation and Transpiration

Evaporation is the transformation of liquid to vapour phase from water, soil and wet surfaces. Sunlight, high air temperature and wind speed accelerates evaporation. On average, evaporation from land surfaces is approximately 60% of the amount of precipitation on the land. Evaporation is measured as the amount of water (as depth) evaporated/ lost from a bucket/pan over a certain time period (usually day or month). The standard measurement practice is that a large metallic pan filled with water is mounted on top of non-metallic supports without any cover. The reduction of water depth in the pan is measured at the end of the measurement period. During the measurement, any precipitation period must be avoided or deducted from the measured loss. Figure 2.13 shows the typical setup of evaporation measurement. As the field measurements are conducted in a metallic pan/bucket, the metal surface gets warmer than the actual water mass, which contributes to a higher evaporation than expected. As such, the actual evaporation from a water surface is smaller than the evaporation measured from a metallic pan. To account for this evaporation measurements are always scaled down using the following equation:

Figure 2.13. Photo showing evaporation measurement
(*Source*: https://www.aboutcivil.org/)

$$E_{actual} = E_{pan} * K \qquad (2.3)$$

where E_{actual} is the actual evaporation, E_{pan} is the evaporation measured at the pan and 'K' is the pan co-efficient ranges 0.64~0.81 depending on the pan material.

2.6.1 Factors Affecting Evaporation

(a) *Vapour pressure of the air above the water surface:* The rate of evaporation is proportional to the difference between the saturation vapour pressure at the water temperature and the actual vapour pressure in the air (above the water surface).
(b) *Water temperature:* If the water is hotter, then its molecules have a higher average kinetic energy. As such, the rate of evaporation will increase with the increase in the water temperature.
(c) *Wind speed:* Gentle wind speed removes the saturated air from the top of the water surface, thus facilitating further evaporation. However, when the wind speed is very high any further increase in wind speed does not increase the rate of evaporation.
(d) *Pressure:* Evaporation happens faster if there is less pressure on the surface from where evaporation is expected (i.e. putting lower force on the molecules to escape). As such, in high altitudes that have a low atmospheric pressure, evaporation will be greater.
(e) *Surface area:* Water having a larger surface area will evaporate faster compared to water of the same volume having a smaller surface area, since there are more surface molecules that are able to escape.
(f) *Density:* The higher the density of water the lower the evaporation rate will be.

2.6.2 Theoretical Calculations of Evaporations

Over the years, several physical and empirical equations were proposed for the calculation of evaporation. Equations are based on commonly available meteorological data. The most suitable equation for a particular locality may be identified through calibration of evaporation measurements for a locality or water body. The following are some widely-used equations for the estimation of evaporation:

(a) Meyer's formula

$$E = K * (e_w - e_a) (1 + u_9/16) \qquad (2.4)$$

where E is the evaporation in mm/day, e_w is the saturated vapour pressure (in mm mercury) for the air above the water, e_a is the actual vapour pressure (in mm mercury) for the air above the water, u_9 is the mean wind velocity (in km/h) at 9 m above the ground. 'K' is the co-efficient incorporating some other factors with values 0.36~0.50.

(b) Rohwer's formula

$$E = 0.771 * (1.465 - .000732 * p_a) * (e_w - e_a) * (0.4 + .0733 * u_0) \quad (2.5)$$

where E, e_w and e_a are as defined earlier; p_a is the mean atmospheric pressure (in mm mercury) and u_0 is the mean wind velocity (in km/h) at ground level (which can be measured at 0.6 m height above the ground).

(c) Penman formula: This equation is based on strong theoretical relations, energy-balance and mass-transfer concepts:

$$E = \frac{mR_n + \rho_a c_p (\delta e) g_a}{\lambda_v (m + \gamma)} \quad (2.6)$$

where E is the evaporation in kg/(m²·s), m is the slope of the saturation vapour pressure curve (in Pa/K), R_n is the net irradiance (W/m²), ρ_a is the density of air (kg/m³), c_p is the heat capacity of air (J/kg/K), g_a is the momentum surface aerodynamic conductance (m/s), δe is the vapor pressure deficit (Pa), λ_v is the latent heat of vaporisation (J/kg) and γ is the psychrometric constant (Pa/K).

2.6.3 Transpiration

Plants extract water from soil; some of this extracted water evaporates through plant leaves/stem/flowers/roots, which is called transpiration. Combined loss of water from both the soil surface and plants is called evapotranspiration. For particular atmospheric conditions, evapotranspiration will depend on the availability of water.

2.7 Catchment and Watershed

It is the area of land that contributes runoff to a specified outlet location. Any rain/water/snowmelt that drops within a point of a catchment will either infiltrate through soil surface or will flow to the outlet as a runoff. The area/boundary of catchment mainly depends on the outlet location. Catchment boundary depends on the land topography and can be identified through contour of the land surface. Catchment area is the plan area (projected onto a horizontal plane) of the delineated catchment boundary. Figure 2.14 shows the catchment boundary of a land at an outlet location (X) drawn off the contour map. For most of the water engineering analysis this is the basic task to be performed, i.e. to calculate the catchment area which will drain to the point of interest. Using digital elevation data, a drafting software like AutoCAD, Civil 3D, 12D can automatically draw the catchment boundary.

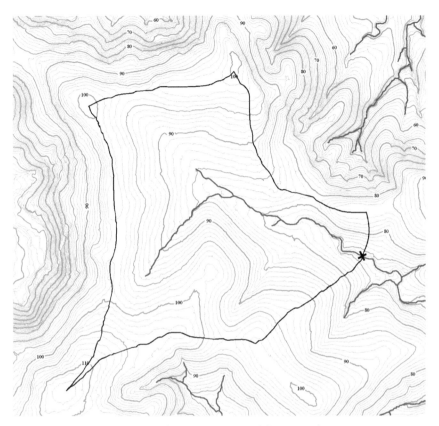

Figure 2.14. An example of catchment area delineation from contour map

Color version at the end of the book

2.8 Abstraction and Losses

Losses/abstractions are generally defined as that part of the rainfall/precipitation that does not show up as runoff. The following are different types of rainfall losses:

(a) **Interception:** The portion of the rainfall that is intercepted by trees, plants, obstacles, and vegetation before it can reach the ground. Interception occurs in the initial part of the storm and eventually the intercepting surfaces become wet (which is their maximum holding capacity).

(b) **Surface depression storage:** The portion of the rainfall prevented from becoming runoff by being trapped in small puddles and depressions on the ground surface. It can occur over pervious and impervious surfaces. The water stored in the depressions will evaporate/infiltrate eventually.

(c) **Infiltration:** This is the flow of water into the ground through the earth's surface. Infiltration rate is a measure of the rate at which soil is able to absorb rainfall or precipitation. It is measured as depth per unit time (i.e. millimeters per hour). For dry soil, the initial rate of infiltration is usually high, however the rate decreases as the soil becomes saturated. When precipitation rate exceeds the infiltration rate, runoff occurs.

The infiltration rate is measured using a simple device named infiltrometer. It is basically two eccentric annular rings/pipes of different diameter penetrated within the soil, as shown in Figure 2.15. Water is poured within the pipe columns and water starts receding within the pipes as infiltration continues. Reductions in water level within the inner pipe are measured at different times until the rate of recession becomes constant. Water filled in the outer pipe helps to eliminate the effect of horizontal flow within the soil. Figure 2.16 shows a typical infiltration rate curve with time for different types of soils, which shows a typical higher infiltration rate at the beginning and becomes constant at one stage. Mathematically, this figure can be represented using Horton's equation:

$$f_t = f_c + (f_0 - f_c) * e^{-kt} \qquad (2.7)$$

where f_t is the infiltration rate at time t; f_0 is the initial infiltration rate, f_c is the constant or equilibrium infiltration rate after the soil becomes saturated or minimum infiltration rate and k is the infiltration decay constant dependent on the soil type.

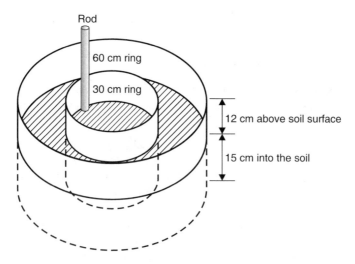

Figure 2.15. A typical infiltration measurement setup
(*Source*: http://www.fao.org/docrep/S8684E/s8684e0a.htm)

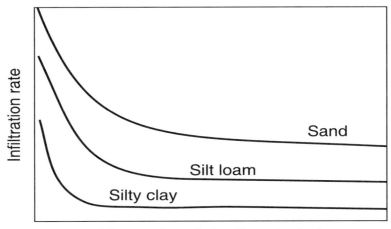

Figure 2.16. Typical infiltration rate curves for different type soils
(*Source*: http://croptechnology.unl.edu/pages/)

2.9 Runoff and Hydrographs

Runoff is the flow of water generated from a catchment through rainfall or snowmelt at a particular location of catchment outlet. Total runoff comprises of surface runoff (generated as overland flow) and groundwater runoff (seepage to channel/stream from groundwater sources). However, the majority of the generated runoff is 'surface runoff', especially in urban settings. As such, for urban catchments, the word 'runoff' usually only refers to the 'surface runoff', which is generated immediately after the rainfall and/or snowmelt. The amount of runoff is greatly influenced by the point of interest (or measurement location), which is termed as 'catchment outlet'. With the change of the outlet location (or point of interest), the contributing area will also change, which will cause the amount of runoff to change significantly. As it is basically the flow of water, it has a unit of flow (i.e. m^3/s, l/s etc.).

A hydrograph is the graphical representation of runoff (or water levels in the stream/river) at a particular point (or catchment outlet). It is a plot of discharge/flow (or water level) versus time at a particular location of a stream/river. Depending on this parameter, hydrographs are categorised as either i) Discharge hydrograph or ii) Stage hydrograph. Figure 2.17 shows a plot of a discharge hydrograph with the associated rainfall hyetograph.

Hydrographs are often shown in one graph with the associated rainfall(s), so that the reader can easily correlate the runoff volume/peak(s) with the rainfall amount/peak(s). From the figure, it is evident that the three runoff peaks are associated with three consecutive rainfall peaks.

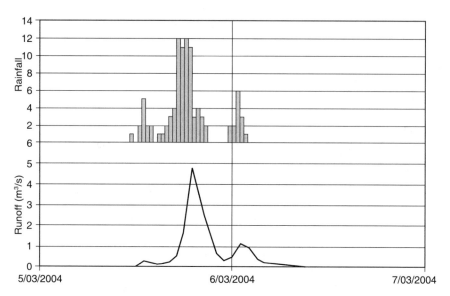

Figure 2.17. A typical hydrograph plot (below) with associated rainfall (above)

However, the ratio of runoff to rainfall for the first peak is much lower than the same ratio for the two subsequent peaks. The reason behind this is that, for the first rainfall peak, the soil was dry/unsaturated; as such, the infiltration loss was higher compared to the two subsequent peaks, when the soil became saturated (or near saturated) and the infiltration rate became very low. With the lower infiltration rate, the ratio of runoff to rainfall is expected to become higher. Runoffs can be calculated (through a theoretically derived equation using measured rainfall and catchment information) and/or measured (using measuring instruments) at a particular point of a stream/river section.

2.10 Streamflow Measurements

For prismatic or small regular laboratory flumes/channels, the flow can be measured directly using flow measuring device(s) or using some weir/gate type structure(s) which will be discussed in another chapter. As for natural channels/streams, flows are measured indirectly; the velocity is measured (with the aid of a velocity meter) at a certain point of the flow, then the measured velocity is multiplied with the area of the flow to work out the discharge/flow. Figure 2.18 shows a photo of a typical velocity/current meter.

As the velocity is not uniform across a stream section, the average velocity of the whole section is considered. Over the years, water professionals through several experiments have simplified the

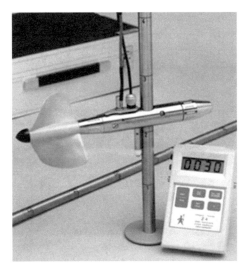

Figure 2.18. A photo of water velocity measuring device
(*Source*: http://www.cleo.net.uk/consultants_resources)

measurement of average velocity using measurement(s) at a minimum number of points. For a single depth measurement (good enough for a shallow depth channel), velocity at 0.6d (i.e. 60% of total depth) can be considered as average depth of the vertical strip (or the whole section if the width is small). For wider streams, several vertical strips (each strip will have separate velocity measurement) should be considered. For a deeper stream, velocities at two different depths (0.2d and 0.8d) should be measured. The average velocity for a deeper stream is worked out using the following equation:

$$V_{av} = \frac{V_{20} + V_{80}}{2} \tag{2.8}$$

where V_{20} is the velocity at 20% depth and V_{80} is the velocity at 80% depth. Figure 2.19 shows a typical stream section having different measurement points (dot points) across the width and depth. The figure also shows the equation to calculate the total discharge of the section through such schematisation.

In recent days, some tiny velocity measuring sensors are also available, however, these sensors are more suitable for precise point measurement of water velocity than for an average bulk water velocity measurement.

Once the flows are measured at different times during/after a rainfall event, the total runoff volume can be estimated and cross-checked with the total amount of rainfall which contributed to that total runoff. Runoff measurements at different times (preferably with equal time steps) are drawn as a runoff hydrograph (Figure 2.20). The total volume of runoff

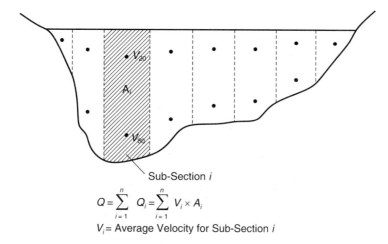

$$Q = \sum_{i=1}^{n} Q_i = \sum_{i=1}^{n} V_i \times A_i$$

V_i = Average Velocity for Sub-Section i

Figure 2.19. A stream sub-sections with measurement points

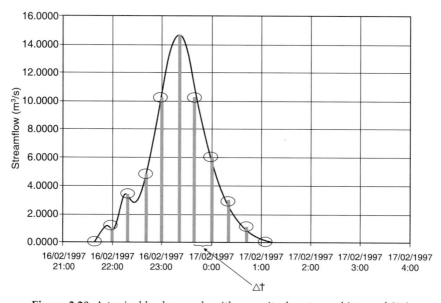

Figure 2.20. A typical hydrograph with magnitudes at equal interval (Δt)

is the total area under this runoff hydrograph. For such estimation, from the different runoff magnitudes (shown as circles) at different time steps, the total volume is calculated using the trapezoidal rule as shown below:

$$\text{Runoff Volume} = \frac{Q_1 + 2 * (Q_2 + \ldots + Q_{L-1}) + Q_L}{2} * \Delta t \qquad (2.9)$$

where Q_1 and Q_L are the first and last magnitudes respectively, Q_2 up to Q_{L-1} are all the intermediate magnitudes, and Δt is the time step.

The total runoff volume can be written in an equation format as below:

Total Runoff Volume = Catchment Area $*\ \Sigma$ (Rainfall – Loss)

Using the above equation, if three parameters are known, the remaining unknown can be estimated. In general, the catchment area is known from the survey, rainfall data is easily obtained from rain gauge measurements. If the total runoff can be estimated from streamflow measurements, the loss can be estimated using this equation.

2.11 Rating Curve

Rating curve is the stage-discharge relationship at a particular section of a stream/river. If the stream/river cross-section and slope remain same, then this relationship would also remain the same. Figures 2.21 and 2.22 show typical rating curves in normal scale and in logarithmic scale, respectively. It is to be noted that, in the rating curves, vertical axis is usually not the depth of water, rather it is the stage (i.e. level or reduced level), which is measured with respect to a certain datum level (Mean Sea Level is usually taken as such datum). In Australia, such datum and measurement standard is called Australian Height Datum (AHD). Every country should have such datum fixed for their referencing for the presentations of elevations. A rating curve can be represented by the following equation:

$$Q - a * (H - b)^n \qquad (2.10)$$

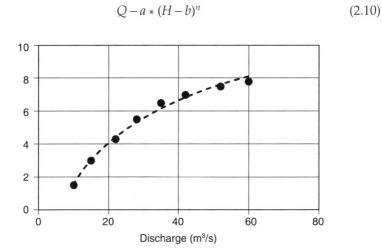

Figure 2.21. A typical stage-discharge rating curve in normal scale

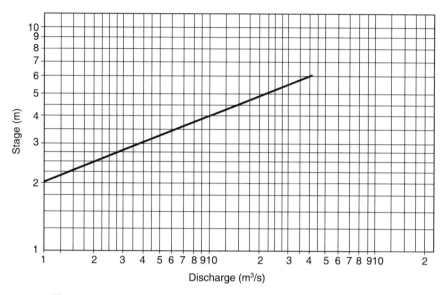

Figure 2.22. A typical stage-discharge rating curve in log-log scale

where Q is the discharge, H is the stage, a and n are rating curve constants and b is the constant representing the gauge reading corresponding to zero discharge.

The main purpose of establishing rating curves is to obtain the discharge measurements easily. Direct discharge measurement is always labour intensive, expensive and an exhaustive task. To avoid these drawbacks, the usual practice is to measure discharge indirectly by using a rating curve. At a particular section of the river/stream, several (preferably having wider range) stage and corresponding discharge measurements are taken, which provide a relationship between discharges and stages at that particular section. From this relationship, in later stages, the discharge can be measured indirectly through direct measurement of the stage, which is easy to measure. In some cases, the curve needs to be extrapolated if the discharge/stage is beyond the measured values. However, discharges estimated from such extrapolated curves may lack accuracy. A change of slope of the straight line in the log-scale rating curve attributes to a sudden change of the stream/river cross-section (i.e. commencement of floodplain from main channel). Such feature is demonstrated in Figure 2.23 and from such figure one should be able to estimate the level of floodplain for the particular river/stream section. From the figure, the floodplain level is approximately equal to 1.7 m.

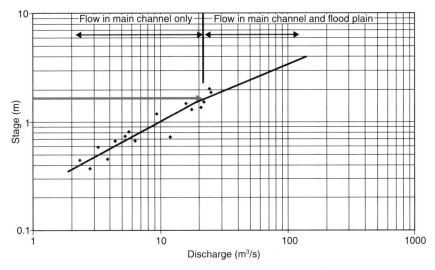

Figure 2.23. A rating curve showing change of slope

Worked Example 1

It is projected that due to global warming, in 2050 the evaporation from ocean surface will increase by 10%, whereas evapotranspiration from land surface will increase by 5%. Consider the current splitting (between land and ocean) ratio of ocean evaporation will remain same as shown in Figure 2.2. Determine the percent increase in precipitation on land.

Solution:

10% increase in ocean evaporation will yield an evaporation flow of 425×1.1 = 467.5 unit, of which evaporation movement to the land would be,

$$467.5/425×40 = 44 \text{ unit.}$$

5% increase in evapotranspiration from land will yield evaporation flow of 71×1.05 = 74.55 unit.

Total moisture precipitating on land = 74.55 + 44 = 118.55 unit

Increase in precipitation = 118.55 – 111 = 7.55 unit

Percent increase in precipitation = 7.55/111×100 = 6.8%

Worked Example 2

A rain gauge has circular-shaped rainwater collection surface with a radius of 40 cm. To increase measuring accuracy (i.e. to be able to measure up to smaller magnitudes) the collection cylinder was narrowed down to a tube having 4 cm radius. A rainwater depth of 10 cm was measured in this narrow tube. What was the actual rainfall depth?

Solution:

Volume of rainwater accumulated in the narrow tube

$$= 10 \times \pi \times 4^2 = 502.4 \text{ cm}^3$$

Original collection area (with radius 40 cm) $= \pi \times 10^2 = 314 \text{ cm}^2$

Now, if the same volume of water spread in to a bucket having surface area of 314 cm^2, depth of water in the bucket = 502.4/314 = 1.6 cm.

So, actual rainfall depth is 1.6 cm.

Worked Example 3

Magnitudes (in mm) of rainfall hyetograph bars are provided below in 15 minutes interval. Convert these magnitudes to a rainfall hyetograph of 30 minutes interval. Also, determine average rainfall intensities (in mm/hr) from this rain event for the whole period, for the first hour and for the first two hours.

$$5, 10, 12, 18, 25, 20, 15, 10, 6, 4$$

Solution:

Magnitudes of hyetograph bars for 30 minutes interval would be just addition of two consecutive values (as each value is for 15 minutes) starting from the beginning. As such, magnitudes of 30 minutes hyetograph bars are:

$$15, 30, 45, 25, 10$$

Total rainfall amount in 2.5 hours = 15+30+45+25+10 = 125 mm and intensity = 125/2.5 = 50 mm/hr

Average intensity during first hour = 15+30 = 45 mm/hr

Total rainfall during first two hours = 15+30+45+25 = 115 mm and average intensity during this two hours = 115/2 = 57.5 mm/hr.

Worked Example 4

There are four rain gauge stations in a large catchment. During a rain event the measured rainfalls in those stations are, 4 mm, 4.5 mm, 5 mm and 6 mm. To be able to use Thiessen mean method, through joining points and drawing perpendicular bisectors, areas associated with the gauges are 6 ha for 4 mm, 5 ha for 4.5 mm, 4 ha for 5 mm and 3 ha for 6 mm. Calculate the average rainfall of the catchment using Thiessen Mean Method.

Solution:

For the catchment,

$$R_1 = 4.0, R_2 = 4.5, R_3 = 5.0, R_4 = 6.0 \text{ in mm, and}$$
$$A_1 = 6.0, A_2 = 5.0, A_3 = 4.0, A_4 = 3.0 \text{ in ha.}$$

Using the Thiessen method equation, average rainfall =

$$(4 \times 6 + 4.5 \times 5 + 5 \times 4 + 6 \times 3)/(6 + 5 + 4 + 3) = 4.69 \text{ mm}$$

Worked Example 5

For a large catchment, shown in Figure 2.12, after a rainfall event (which occurred with variable depths over the catchment) contours of rainfall depths were drawn at 1 mm interval. The drawn contours divided the whole catchment in to three sub-areas and the sub-areas were measured (in ha) as given below:

$$A_1 = 4.0, A_2 = 5.0, A_3 = 6.0;$$

Rainfall depth (in mm) contour values are:

$$R_1 = 2.0, R_2 = 3, R_3 = 4, R_4 = 5.$$

Determine the average rainfall of the catchment for the rain event using isohyetal method.

Solution:

Using isohyetal method equation, average rainfall =

$$[4 \times (2+3)/2 + 5 \times (3+4)/2 + 6 \times (4+5)/2]/(4+5+6) = 3.63 \text{ mm}$$

Worked Example 6

A new evaporation tank is being used for evaporation measurement with unknown pan coefficient. For the purpose of calibration, the tank is placed near to another evaporation tank whose pan coefficient (K) is known to be "0.75". During a particular day monitoring, the new tank had an evaporation reading of 11 mm, while the old tank had a loss (i.e. evaporation) of 12 mm. Determine the actual evaporation and the pan coefficient for the new tank.

Solution:

Actual evaporation = $E_{pan} \times K = 12 \times 0.75 = 9$ mm

Now, pan coefficient for the new tank = (Actual evaporation)/E_{pan} = $9/11 = 0.818$.

Worked Example 7

For an infiltration rate test, the following parameters were measured:

Initial infiltration rate = 12 mm/hr, Final infiltration rate = 2 mm/hr, Infiltration rate after 3 days = 8 mm/hr.

Determine the recession co-efficient of the soil on which test was conducted.

Solution:

For the above data, in the infiltration equation, $f_t = f_c + (f_0 - f_c) * e^{-kt}$

$$f_0 = 12, f_c = 2 \text{ and } f_3 = 8, \text{ where, } t = 3$$

Using the above equation,

$$8 = 2 + (12 - 2) * e^{-k*3}$$

$$8 - 2 = 10 * e^{-k*3}$$

$$e^{-k*3} = 6/10 = 0.6$$

Then taking ln in both sides,

$$-k * 3 = \ln(0.6) = -0.511$$

Thus, $k = 0.511/3 = 0.17$
So, recession co-efficient of the soil is 0.17.

Worked Example 8

For an infiltration rate test, the following parameters were measured:
Final infiltration rate = 2 mm/hr, Infiltration rate after 1 day = 12 mm/hr, Infiltration rate after 2 days = 10 mm/hr. Determine the recession co-efficient of the soil on which test was conducted. Also, calculate the initial infiltration rate for the same testing.

Solution:

For the above data, the infiltration equation, $f_t = f_c + (f_0 - f_c) * e^{-kt}$ has to be written twice for two data sets (i.e. $t = 1$ day and $t = 2$ days). Also, given final infiltration rate, $f_c = 2$.

For $t = 1$, $12 = 2 + (f_0 - 2) * e^{-k*1}$,

which can be written as:

$$12 - 2 = 10 = (f_0 - 2) * e^{-k*1} \tag{1}$$

For $t = 2$, $10 = 2 + (f_0 - 2) * e^{-k*2}$,

which can be written as:

$$10 - 2 = 8 = (f_0 - 2) * e^{-k*2} \tag{2}$$

Now, diving Equation 1 by Equation 2,

$$\frac{10}{8} = \frac{(f_0 - 2) * e^{-k*1}}{(f_0 - 2) * e^{-k*2}} = e^{2k-k} = e^k$$

So, $e^k = 1.25$, taking ln in both sides yield,

$$k = \ln(1.25) = 0.223$$

So, recession co-efficient is 0.223.

Now, replacing this value of 'k' in to Equation 1,

$$12 = 2 + (f_0 - 2) * e^{-0.223}, \text{ which yields}$$

$$(f_0 - 2) = 10/e^{-0.223} = 10 * e^{0.223} = 12.50 \text{ and}$$

$$f_0 = 12.50 + 2.0 = 14.50$$

So, initial infiltration rate was 14.5 mm/hr.

Worked Example 9

At a creek cross-section, velocity measurements were taken. The cross-section was divided into five sub-sections, each section having different velocity measurements at different depths (d), as shown in the table below. Area of each sub-section is also provided in the table. From the measurements, calculate the total discharge through the creek cross-section.

		Section 1	Section 2	Section 3	Section 4	Section 5
Area (m^2)		0.20	0.50	0.60	0.30	0.10
Velocity at different depths (cm/s)	0.2d	8	25	30		
	0.6d	15			24	10
	0.8d		30	40		5

Solution:

Note: For this problem two things should be noticed; 1) The unit of area is in square meters, whereas the unit of velocity is in cm/s, so conversion of one unit is needed and 2) In Sections 1 and 5, in addition to velocity measurements being given, which are not useful in getting the average velocity from 0.2d and 0.8d measurements, we need both the measurements at 0.2d and 0.8d.

Average velocities for all the sections:

$$V_1 = 15 \text{ cm/s} = 0.15 \text{ m/s}$$

$$V_2 = (25 + 30)/2 = 27.5 \text{ cm/s} = 0.275 \text{ m/s}$$

$$V_3 = (30 + 40)/2 = 35 \text{ cm/s} = 0.35 \text{ m/s}$$

$$V_4 = 24 \text{ cm/s} = 0.24 \text{ m/s}$$

$$V_5 = 10 \text{ cm/s} = 0.10 \text{ m/s}$$

Now, total discharge,

$$Q = \Sigma\, A_i V_i = A_1 {*} V_1 + A_2 {*} V_2 + A_3 {*} V_3 + A_4 {*} V_4 + A_5 {*} V_5$$

$$= 0.2{*}0.15{+}0.5{*}0.275{+}0.6{*}0.35{+}0.3{*}0.24{+}0.1{*}0.1$$

$$= 0.4595 \text{ m}^3/\text{s}.$$

Worked Example 10

Stage-discharge data for a particular river section is provided in the table below. The river cross-section has got two floodplains at two different levels. From the provided data, after drawing the stage-discharge relationship in a log-log plot, find the AHD (Australian Height Datum) levels of the lower and upper floodplains.

AHD Level (m)	Discharge (m³/s)
1.75	2.0
2.50	5.0
3.30	10.0
4.0	17.0
5.4	60.0
5.7	100.0

Solution:

Through plotting the values given above (discharge values in x-axis and level values in y-axis) in a log-log plot, the following graph is prepared:

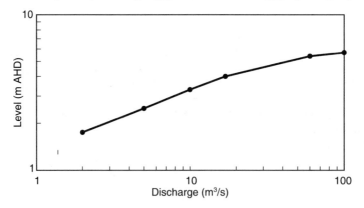

From the above-graph, first change of slope occurs at level 4.0 m and the next change of slope occurs at level 5.4 m. So, level of lower floodplain is 4.0 m AHD and level of higher floodplain is 5.4 m AHD.

Worked Example 11

Table 1 below shows the hyetograph of rainfall recorded at a rainfall station located in a 5.00 km^2 catchment. At a stream gauging station at the catchment outlet, runoff was measured during the rainfall event and presented in Table 2 and Figure 1. Calculate the rainfall excess (in mm) and the losses (both in mm and as a percentage of rainfall).

Table 1

Time (hr)	1	2	3	4	5
Rain (mm)	10	15	25	20	5

Table 2

Time (hr)	1	3	4	5	7
Runoff (m^3/s)	0	10~15	20	15~10	0

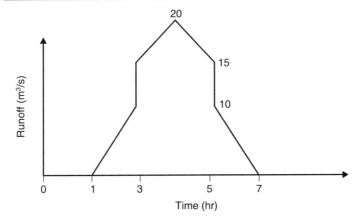

Figure 1

Solution:

Total rainfall, $R = 10 + 15 + 25 + 20 + 5 = 75$ mm

Area, $A = 5$ km^2

Runoff volume = Area of the hydrograph

(As the hydrograph is symmetrical, we can calculate only one half and multiply it by 2 to get the total area)

So, area of the hydrograph = 2*[0.5*(3–1)*10 + 0.5*(15+20)*(4–3)]*60*60
= 198,000 m^3 = Total runoff volume

So, rainfall excess for the rainfall event, RE = Total runoff volume/ Area = $198000/(5*10^6)$ = 0.0396 m = 39.6 mm

So, rainfall loss = $R - RE$ = 75 − 39.6 = 35.4 mm

Loss (in %) = 35.4/75*100 = 47.2%

Worked Example 12

Table 1 below shows the hyetograph of rainfall recorded at a rainfall station located in a 5.00 km^2 catchment. At a stream gauging station at the catchment outlet, runoff was measured during the rainfall event and presented in Table 2 and Figure 1. You may have to use the 'Trapezoidal rule' to calculate total runoff volume. Calculate the rainfall excess (in mm) and the losses (both in mm and as a percentage of rainfall).

Table 1

Time (hr)	Rain (mm)
1	10
2	15
3	25
4	20
5	5
6	5

Table 2

Time (hr)	Runoff (m^3/s)
0	0
1	3
2	8
3	14
4	20
5	14
6	8
7	3
8	0

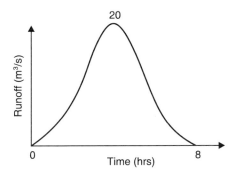

Figure 1

Solution:

Total rainfall, $R = 10+15+25+20+5+5 = 80$ mm

Area, $A = 5$ km^2

Runoff volume = Area of the hydrograph
　　　Applying trapezoidal rule, area of the hydrograph

$$= [0+0+2*(3+8+14+20+14+8+3)]/2*60*60$$

$$= 252,000 \text{ m}^3 = \text{Total runoff volume}$$

So, rainfall excess for the rainfall event, RE = Total runoff volume/ Area $= 252000/(5*10^6) = 0.504 = 50.4$ mm

So, rainfall loss $= R - RE = 80 - 50.4 = 29.6$ mm

Loss (in %) $= 29.6/80*100 = 37\%$

Worked Example 13

The table below shows the runoff hydrograph from a particular rain event drained from a catchment. After analysis, it was found that the rainfall excess for the corresponding rainfall event was 13.32 cm. Calculate the area of the catchment.

Table

Time (hr)	0	2	4	6	8	10	12	14
Runoff (m³/s)	0	4	6	12	6	4	3	0

Solution:

Applying trapezoidal rule, total runoff volume (i.e. area of hydrograph) = $[0+2+2*(4+6+12+6+4+3)]/2*(2*60*60) = 259,200$ m^3

　　Given, rainfall excess (RE) = 13.32 cm = 0.1332 m

Runoff Volume = $RE*$Area

So, Area = Runoff volume$/RE$ = 259200/0.1332 = 1945946 m^2 = 1.946 km^2.

Worked Example 14

Stage-discharge data for a particular river section is provided in the table below. From the provided data, after drawing the stage-discharge relationship in a normal plot, derive the equation of the rating curve as per the format provided in Section 2.11. The level corresponding to zero discharge is 4.0 m AHD.

AHD Level (m)	Discharge (m³/s)
5.75	2.0
6.5	5.0
7.3	10.0
8	17.0
10.4	60.0
11.5	100.0

Solution:

Recommended format of the equation is,

$$Q = a * (H - b)^n$$

The 'b' value is given as 4.0 m. As such, we need arrange the data as $(H - b)$ vs. Q (i.e. discharge) as follows:

(H − b) in m	Discharge (m³/s)
1.75	2.0
2.5	5.0
3.3	10.0
4	17.0
6.4	60.0
7.5	100.0

Now, drawing $(H - b)$ versus 'Q' values produces a graph as shown below:

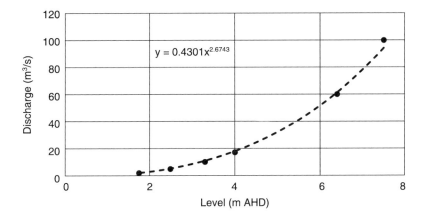

After inserting the data in the Excel sheet, from the Excel function "Add trendline" and selecting the "Power function" will provide a best-fit line, as shown above. Then, selecting "Display equation on the chart" will provide the equation, as shown on the chart, where 'y' is the discharge and 'x' is the $(H - b)$ values, and the derived rating curve is:

$$Q = 0.4301 * (H - 4.0)^{2.6743}$$

Reference

Chow, V.T., Maidment, D.R. and Mays, L.W. (2017). Applied Hydrology. McGraw Hill India, ISBN-10: 9780070702424.

Probabilistic Rainfall/
Flood Estimation

3.1 Introduction to Flood Estimation

Flood estimation is one of the major tasks that Civil/Water engineers deal with. Rainfall amount and pattern are the main focuses for hydrologists and meteorologists. However, the subsequent generation of runoff/flood from the catchment is the main focus for Civil/Water engineers. Primary objectives of flood estimation are:

(a) To determine whether a riverbank/floodplain will be inundated or not
(b) To calculate the depth of flooding in the surrounding neighbourhood/ floodplain
(c) To estimate the potential impact of flooding in the surrounding area
(d) To generate flood prediction/warning for the community
(e) To prepare flood-policy and/or flood-response plans for the residents and authority

In general, there are two types of flood estimations:

(a) Probabilistic or statistical estimation – this is to provide a probability of occurrence of a flood/rainfall having a certain magnitude. It is conducted through statistical analysis of observed floods/rainfalls from long period.
(b) Deterministic flood estimation – this is to calculate the exact flood (i.e. discharge) magnitude from a certain rainfall amount/event. It is performed through calculating the discharge from a certain catchment area due to a certain rainfall amount.

3.2 Terminologies used in Probability Analysis

For the purpose of performing probability analysis, the term 'Average Recurrence Interval (ARI)' is commonly used. ARI (also known as return period) is the average interval of time between events (i.e. rainfall, flood) of specified magnitude (or more) for a certain location. To explain it more clearly, suppose for a particular area, ARI of 100 mm rainfall (in 24 hours) is 10 years. This implies that on average in that area, a rainfall of magnitude 100 mm or more (in 24 hours) is expected to occur once in 10 years or 5 times in 50 years. Another term, 'Exceedances per Year (EY)' is also recently being used mainly for frequent events. EY is basically the number of occurrences of an event having a specified magnitude or more in a year. For example, if, in a certain city, it is usual to have a rainfall magnitude (in a day) equal to or more than 50 mm twice in a year. So, EY of a rainfall intensity 50 mm/day for that city would be "2". An "EY = 2" is equivalent to a 6 months' recurrence interval. Basically, EY is the inverse of ARI, as shown below:

$$EY = 1/ARI \qquad (3.1)$$

ARI also can be explained in an indirect reverse way by using a probabilistic approach. If the probability of occurrence of an event having a magnitude equal to or more than a specified value is 'P', then recurrence interval or return period (T) can be defined as:

$$T = 1/P \qquad (3.2)$$

Another general probabilistic term named 'Annual Exceedance Probability (AEP)' is commonly used for this purpose. AEP is the probability that an event magnitude is exceeded once (or more than once) in a year. An 'AEP' value of '0' implies that the event is never expected to occur and, on the other hand, an 'AEP' value of '1' implies that the event is always expected to occur. The city of Melbourne, for example, receives an average annual rainfall amount of 640 mm and a rainfall amount of 500 mm in a day has never happened. So, for a hypothesis that 'in a year' Melbourne will receive total rainfall amount of 20 mm (or more), 'AEP' for this would be '1' (i.e. will certainly happen). On the other hand, for a hypothesis that 'in a day' Melbourne will experience a total rainfall amount of 500 mm (or more), 'AEP' for this would be '0' (i.e. will never occur). AEPs can be expressed in two different ways; a) AEP expressed as % and b) AEP expressed as '1 in X years'. For instance, an AEP of '1 in 100 years' is equal to 1% AEP and an AEP of '1 in 20 years' is equal to 5% AEP. AEP and EY can be correlated with the following equation (Ball et al., 2016):

$$EY = -\log_e \{1 - AEP\ (\%)\} \qquad (3.3)$$

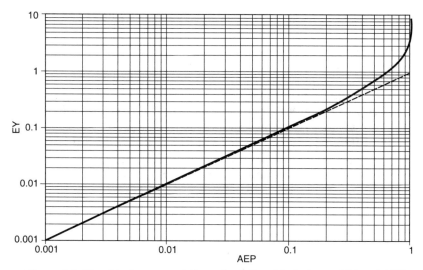

Figure 3.1. Graphical representation of relationship between EY and AEP
(*source*: Ball et al., 2016)

In this equation, for 50% AEP, we have to insert 0.50 and for 1% AEP, we have to insert 0.01. Figure 3.1 shows the graphical representation of the above equation.

Table 3.1 shows the inter-relations among EY, AEPs and ARI as presented by the Australian Rainfall and Runoff (http://arr.ga.gov.au/). In the table, the shaded cells are 'acceptable terminologies' and the thick outlined cells are 'preferred terminologies' under different category of rain events (i.e. very frequent, frequent, rare, very rare and extreme) by Australian authority. As depicted in the table, Australian authorities no longer prefer to use 'ARI' terminology, however many other countries still use 'ARI' and the basic concept is same for other terminologies. Figure 3.2 shows the general trend of rainfall magnitudes with respect to EYs and AEPs. Although, in Figure 3.2, "Design Rainfall Depth" is mentioned on the vertical axis, similar trend is expected to be observed for other hydrological variables, such as discharge/flood, evaporation, wind speed and high temperature. Figure 3.3 show typical relationships of flood magnitudes with ARI, AEP (1 in X), AEP (%) and EY. It is obvious that for higher ARIs, magnitudes of rainfall/flood will be also higher. As, for example, for a particular location if 1 year ARI (for a particular time period) rainfall intensity is 20 mm/hr, then 100 year ARI rainfall intensity for the same location would be much higher.

3.3 Failure and Risk

In hydraulics (especially in flood and drainage studies), failure does

Table 3.1. Inter-relations among EY, AEPs and ARI

Frequency descriptor	EY	AEP <%)	AEP (1 in x)	ARI
Very Frequent	12			
	6	99.75	1.002	0.17
	4	98.17	1.02	0.25
	3	95.02	1.05	0.33
	2	86.47	1.16	0.5
	1	63.21	1.58	1
Frequent	0.69	50	2	1.44
	0.5	39.35	2.54	2
	0.22	20	5	4.48
	0.2	18.13	5.52	5
	0.11	10	10	9.49
Rare	0.05	5	20	20
	0.02	2	50	50
	0.01	1	100	100
Very Rare	0.005	0.5	200	200
	0.002	0.2	500	500
	0.001	0.1	1000	1000
	0.0005	0.05	2000	2000
Extreme	0.0002	0.02	5000	5000
			↓	
			PMP/ PMPDF	

Source: Ball et al. (2016)
PMP: Probably Maximum Precipitation

not necessarily mean collapse of a relevant structure. As, for example, floodwater overflowing the bridge deck during a flood event, this can be termed as hydraulic failure of the bridge, however the bridge might be still intact. Hydraulic structures are designed for a specific design flow (i.e. frequency of exceedance), hence, when that design flow exceeds, the

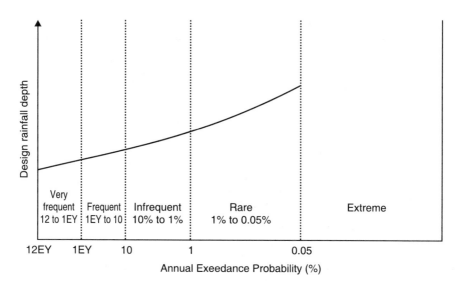

Figure 3.2. Trend of rainfall magnitudes with EYs and AEPs
(*Source*: Ball et al., 2016)

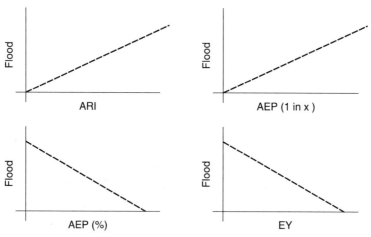

Figure 3.3. Typical relationships of flood magnitudes with ARI, AEP
(1 in X), AEP (%) and EY

structure is likely to be overflowing/flooding, which can be termed as failure.

Surcharging is another term of failure. As mentioned earlier, hydraulic structures are designed to carry a certain design flow, however to avoid/ minimise the loss/damage during a flood having higher flow, this is customary to provide a safe passage of such overflow/surcharge. Such provision of safe passage is often termed as 'overflow bypass'.

In regard to risk, it is very difficult to quantify and the community lacks understanding about it. Risks are represented through probabilistic terminologies, as explained in the earlier section. Also, the acceptable level of risk depends on service level acceptable by the community, which largely varies with the country and social status of the community. As for example a developing country may accept a higher level of risk, while a wealthy country will try to avoid even the lower level of risks.

In general, if monetary cost (or social damage) of a failure is small, then frequent failures may be acceptable. For example, flooding/surcharging of a street kerb is not likely to cause a major economic/social damage. As such, frequent flooding of a street kerb may be acceptable to some extent. Depending on the community/locality it may be acceptable that street kerbs are flooded 1~4 times in a year. If the street is in front of a school/childcare centre/hospital, the level of acceptance may likely be 'flooding once in a year', depending on affordability. However, for other general streets not having any high-profile nearby structures, such acceptance may likely to be 4 times (or even more) per year. Such structures are often termed as 'minor structures'. However, for a 'major structure', where cost of damage (economically and socially) is large, frequent failure is not acceptable. For example, risk of a house floor level getting flooded, the community may accept a frequency of hundred years or more depending on affordability. However, in regard to only house premise (front yard or backyard) flooding, the community may accept more frequent floods.

The following are some recommended AEP events for the selection of design flood:

- A culvert crossing a rural road with low traffic flow: 1 in 1 year AEP
- A street kerb along an urban road: 1 in 1 year AEP
- A drainage pipe running along an urban road or culvert crossing an urban road with medium traffic flow: 1 in 5 years to 1 in 10 years AEP
- An urban road with high traffic flow: 1 in 20 years to 1 in 50 years AEP
- A bridge connecting to a suburb: 1 in 50 years to 1 in 100 years AEP
- A Bridge connecting to a Hospital/Childcare/School: >> 1 in 100 years AEP

In general, a flood event having 1 in 100 years AEP is considered as a major event. Such major flood magnitude can be used for different purposes for the design of different level of structures.

- For small structures, such high magnitude flood is used to check the scenario, i.e. what will happen if such a flood occurs. However, such magnitude flood is not used for the design of small structures
- For medium structures, such high magnitude flood is used for the sizing of some major components of the structure (i.e. not for the full structure). Also, such high magnitude flood is sued for scenario study.

- For large structures, high magnitude flood is used for the sizing for the structure (i.e. the structure should easily carry/pass such high discharge).

3.4 Hydrological Data

For hydrological and flood analysis the following data are required:
- Rainfall data
- Evaporation data
- Imperviousness of the surface
- Vegetation type
- Soil properties (including moisture content)
- Discharge/streamflow data

Data is often limited, which hinders a complete and accurate flood study. Also, in some cases the quality of collected data is questionable. Deterministic flood estimations require more data compared to probabilistic estimations. As such probabilistic flood estimations are widely used in many countries.

3.5 Flood Frequency Analysis

Flood frequency analysis is basically a statistical analysis of observed/measured flood discharges from at least several years. It determines the relationship between ARI/AEP and flood magnitudes. For analysis to be valid or accurate, the data should constitute samples of independent values from homogenous population. Suppose a dam is located upstream of a stream/river section and, as such, flow is regulated by the authority based on different needs. So, a flow measured in that section of the river is not natural/independent. Flood frequency analysis should not be done with such data. In regard to data homogeneity, it may occur due to the following reasons:

- Extrapolated values of discharges from rating curve: Often discharge is measured from rating curve, extrapolated values from rating curve (especially during extreme events) do not always provide correct discharge values.
- Failure of gauge during extreme events: Often, during extreme events, gauges are displaced/removed by the flood. In such cases, collected data is either missing or erroneous.
- Change of gauge site: Sometimes it becomes necessary to change the gauge location due to other river training works or erosion of the gauge site. In such case, if the replaced gauge location is far from the previous location, the collected data from the new location will not be homogenous to the earlier data.

- Construction of hydraulic structure near the gauge site at a later period: Due to construction of hydraulic structures (bridge, culvert, sluice gate, weir, etc.), water level-discharge characteristics of the gauge site are altered.
- Change in land use characteristics of the catchment: Suppose 50 years discharge data is available for a river section, which is draining from a catchment. However, during the last 10 years, the contributing catchment has gone through significant changes due to urbanisation (or deforestation). In this case, the last 10 years of data is not homogenous with the first 40 years of data.

There are two commonly adopted methods for flood frequency analysis. These methods can be also used for other natural events, such as rainfall, high temperature, wind speed, bushfire, snowfall, hurricane, etc. The methods are described in the following sections.

3.5.1 Annual Maximum Series Analysis

This method analyses maximum discharge in each year of record from a particular river/stream section. It estimates the probability that a maximum flood discharge may exceed a particular magnitude in a year, or in other words, it is the average period between years where the specified maximum discharge is exceeded once. In short, it is also called "Annual Series Analysis". To perform this analysis, it requires the extraction of maximum discharge from each year from a historical record of many years. It is to be noted that the considered year can comprise of months from January to December, or water year (as it is practiced in some countries, and not starting from January). For such analysis, the number of years for the analysis is equal to the records available for the analysis. For example, Table 3.2 shows several flood estimates for the mentioned years.

Table 3.2. Recorded floods in different years at a river section

Year	2000	2001	2002	2003	2004
Flood (m^3/s)	50, 85, 64, 106	68, 73, 110, 77	89, 74, 66, 50	86, 78, 55, 65	51, 95, 88, 75

From the above recorded data, if only the maximum value from each year is taken, then the extracted maximum flood magnitudes are shown in Table 3.3.

Table 3.3. Recorded maximum flood in each year at the river section

Year	2000	2001	2002	2003	2004
Flood (m^3/s)	106	110	89	86	95

Therefore, further analysis will be conducted using only the values shown in the latest table. Analysis can be performed through two ways; a) Graphical method and b) Analytical method. However, the final results may vary among these methods.

Graphical Method

This method is basically to draw a relationship graph (as shown in Figure 3.4) between AEPs and flood discharges based on the given historical flood data. Then, from the drawn graph, flood magnitude corresponding to any particular AEP can be extracted. Alternatively, for any flood discharge, the corresponding AEP value can be extracted. The same graph can be used to find rare AEP flood magnitudes through extrapolation of the graph in the case of limited data availability. For example, a frequency analysis is conducted with twenty years of recorded annual maximum floods. The Graph produced from such dataset will not cover a 1% AEP value, however such a graph can be extrapolated to extract a flood magnitude having 1% AEP.

Procedure of graphical annual series analysis: Flood magnitudes of the recorded years are ranked for each flood event, as '$m = 1$' (for the biggest flood) and '$m = N$' (for the smallest flood), where 'N' is the number of years of record. Then statistical 'Plotting Position (PP)' (which is equivalent to AEP) is calculated for each of the flood events, as per the following equation:

$$\text{AEP} = \frac{m - 0.4}{N + 0.2} \qquad (3.4)$$

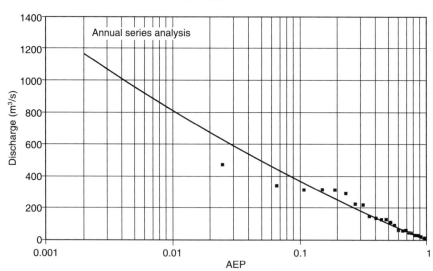

Figure 3.4. Typical AEP versus discharge graph
(not representing the data mentioned in this section)

It is to be noted here that there are several different but similar equations proposed by different scientists and each of those equations may be suitable for a particular variable (i.e. temperature, earthquake, wind speed etc.) or locality. Then, the calculated AEP values are plotted against corresponding flood discharge values in a semi-log paper (discharge values in normal scale and AEP values in log scale). Finally, a best-fit curve (usually straight line) is drawn through the points and this best-fit curve is the relationship between AEPs and flood discharges, which can later be used for the extraction of AEP (for a particular flood magnitude) or flood magnitude for a particular AEP.

Analytical Method

This method is basically used to derive an equation representing the relationship between AEPs and flood magnitudes, with drawing/using the above-mentioned graph/plot. Steps to be followed for the analytical method are outlined below:

Step 1: Calculate the log (base 10) for each discharge, X_i

Step 2: Calculate mean (M) for all the (log X_i) values

Step 3: Using Equation 3.5 in order to calculate standard deviation (S) of all the (log X_i) values

Step 4: Using Equation 3.6, calculate the coefficient of skewness (g) of all the (log X_i) values

Step 5: Find Frequency Factors (K_y) based on 'g' for a particular AEP from Table 3.4

Step 6: Calculate discharge (Q_y) for any particular AEP (y) using Equation 3.7

$$S = \sqrt{\frac{\Sigma(X_i - M)^2}{N}} \qquad (3.5)$$

$$g = \sqrt{\frac{\Sigma(X_i - M)^3}{(N-1)*S^3}} \qquad (3.6)$$

$$\log Q_y = M + K_y S \qquad (3.7)$$

3.5.2 Peak-Over-Threshold Series Analysis

Similar to the 'graphical method' of the 'Annual Maximum Series Analysis', this method draws a relationship graph (as shown in Figure 3.5) between ARIs and flood discharges based on the given historical flood data. Then,

Table 3.4. Frequency factors *Kv* for use with Log-Pearson Type III distribution – From Australian Rainfall and Runoff (1998)

	ARI (Y years)											
	500	200	100	50	20	10	5	2	1.25	1.1111	1.0526	1.0101
	Probability (Percent)											
Skew	0.2	0.5	1	2	5	10	20	50	80	90	95	99
-2.0	0.998	0.995	0.990	0.980	0.949	0.895	0.777	0.307	-0.609	-1.303	-1.996	-3.605
-1.9	1.049	1.044	1.037	1.023	0.984	0.920	0.788	0.294	-0.627	-1.311	-1.989	-3.553
-1.8	1.105	1.097	1.087	1.069	1.020	0.945	0.799	0.282	-0.643	-1.318	-1.981	-3.499
-1.7	1.165	1.155	1.140	1.116	1.056	0.970	0.808	0.268	-0.660	-1.324	-1.972	-3.444
-1.6	1.231	1.216	1.197	1.166	1.093	0.994	0.817	0.254	-0.675	-1.329	-1.962	-3.388
-1.5	1.303	1.282	1.256	1.217	1.131	1.018	0.825	0.240	-0.691	-1.333	-1.951	-3.330
-1.4	1.380	1.351	1.318	1.270	1.168	1.041	0.832	0.225	-0.705	-1.337	-1.938	-3.271
-1.3	1.462	1.424	1.383	1.324	1.206	1.064	0.838	0.210	-0.719	-1.339	-1.925	-3.211
-1.2	1.550	1.501	1.449	1.379	1.243	1.086	0.844	0.195	-0.733	-1.340	-1.910	-3.149
-1.1	1.643	1.581	1.518	1.435	1.280	1.107	0.848	0180	-0.745	-1.341	-1.894	-3.087
-1.0	1.741	1.664	1.588	1.492	1.317	1.128	0.852	0.164	-0.758	-1.340	-1.877	-3.023
-0.9	1.842	1.749	1.660	1.549	1.353	1.147	0.854	0.148	-0.769	-1.339	-1.859	-2.957

(Contd.)

-0.8	-2.891	-1.839	-1.336	-0.780	0.132	0.856	1.166	1.389	1.606	1.733	1.837	1.948
-0.7	-2.824	-1.819	-1.333	-0.790	0.116	0.857	1.183	1.423	1.663	1.806	1.926	2.057
-0.6	-2.755	-1.797	-1.329	-0.800	0.099	0.857	1.200	1.458	1.720	1.880	2.016	2.169
-0.5	-2.686	-1.774	-1.323	-0.808	0.083	0.857	1.216	1.491	1.777	1.955	2.108	2.283
-0.4	-2.615	-1.750	-1.317	-0.816	0.067	0.855	1.231	1.524	1.834	2.029	2.201	2.399
-0.3	-2.544	-1.726	-1.309	-0.824	0.050	0.853	1.245	1.555	1.890	2.104	2.294	2.517
-0.2	-2.472	-1.700	-1.301	-0.830	0.033	0.850	1.258	1.586	1.945	2.178	2.388	2.637
-0.1	-2.400	-1.673	-1.292	-0.836	0.017	0.846	1.270	1.616	2.000	2.253	2.482	2.757
0.0	-2.326	-1.645	-1.282	-0.842	0.000	0.842	1.282	1.645	2.054	2.326	2.576	2.878
0.1	-2.253	-1.616	-1.270	-0.846	-0.017	0.836	1.292	1.673	2.107	2.400	2.670	3.000
02	-2.178	-1.586	-1.258	-0.850	-0.033	0.830	1.301	1.700	2.159	2.472	2.763	3.122
0.3	-2.104	-1.555	-1.245	-0.853	-0.050	0.824	1.309	1.726	2.211	2.544	2.856	3.244
0.4	-2.029	-1.524	-1.231	-0.855	-0.067	0.816	1.317	1.750	2.261	2.615	2.949	3.366
0.5	-1.955	-1.491	-1.216	-0.857	-0.083	0.808	1.323	1.774	2.311	2.686	3.041	3.487
0.6	-1.880	-1.458	-1.200	-0.857	-0.099	0.800	1.329	1.797	2.359	2.755	3.132	3.609
0.7	-1.806	-1.423	-1.183	-0.857	-0.116	0.790	1.333	1.819	2.407	2.824	3.223	3.730
0.8	-1.733	-1.389	-1.166	-0.856	-0.132	0.780	1.336	1.839	2.453	2.891	3.312	3.850

(Contd.)

Table 3.4. (*Contd.*)

Skew	1.0101	1.0526	1.1111	1.25	2	5	10	20	50	100	200	500
						ARI (Y years)						
					Probability (Percent)							
0.9	-1.660	-1.353	-1.147	-0.854	-0.148	0.769	1.339	1.859	2.498	2.957	3.401	3.969
1.0	-1.588	-1.317	-1.128	-0.852	-0.164	0.758	1.340	1.877	2.542	3.023	3.489	4.088
1.1	-1.518	-1.280	-1.107	-0.848	-0.180	0.745	1.341	1.894	2.585	3.087	3.575	4.206
1.2	-1.449	-1.243	-1.086	-0.844	-0.195	0.733	1.340	1.910	2.626	3.149	3.661	4.323
1.3	-1.383	-1.206	-1.064	-0.838	-0.210	0.719	1.339	1.925	2.667	3.211	3.745	4.438
1.4	-1.318	-1.168	-1.041	-0.832	-0.225	0.705	1.337	1.938	2.706	3.271	3.828	4.553
1.5	-1.256	-1.131	-1.018	-0.825	-0.240	0.691	1.333	1.951	2.743	3.330	3.910	4667
1.6	-1.197	-1.093	-0.994	-0.817	-0.254	0.675	1.329	1.962	2.780	3.388	3.990	4.779
17	-1.140	-1.056	-0.970	-0.808	-0.268	0.660	1.324	1.972	2.815	3.444	4069	4890
1.8	-1.087	-1.020	-0.945	-0.799	-0.282	0.643	1.318	1.981	2.848	3.499	4.147	4.999
1.9	-1.037	-0.984	-0.920	-0.788	-0.294	0.627	1.311	1.989	2.881	3.553	4.223	5.108
2.0	-0.990	-0.949	-0.895	-0.777	-0.307	0.609	1.303	1.996	2.912	3.605	4.298	5.215

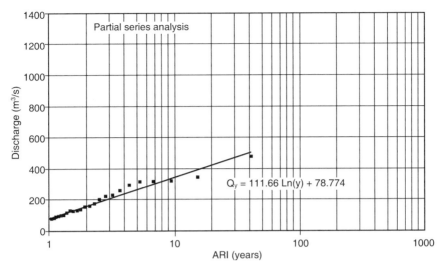

Figure 3.5. Typical ARI versus discharge graph (not representing the data mentioned in this section)

from the graph, the flood magnitude corresponding to any particular ARI can be extracted. Alternatively, for any flood discharge, the corresponding ARI value can be extracted. Also, the drawn graph can be extrapolated in order to extract flood magnitudes of extreme events (i.e. very high ARIs). The Only difference from the 'Annual Maximum Series Analysis' is that in the 'Annual Series Analysis' only the maximum flood magnitude from each year is considered, whereas in this method all the flood magnitudes above (or equal to) a selected threshold value are considered. Usually, 'threshold value' is selected by the authority/stakeholder in a way that at least one flood value is selected from each year. As for example, from Table 3.1 any flood magnitude of "86" or lower can be selected as threshold value, however any value above "86" should not be considered as the threshold value, as in such case the data of the year 2003 will be missed out. As all the flood magnitudes above the threshold value is considered, the number of records are usually more than the number of considered years (N). This method is also termed 'Partial Series Analysis'.

Procedure of graphical partial series analysis: Flood magnitudes of the recorded years are ranked for each flood event, as '$m = 1$' (for the biggest flood). In this case, for the smallest flood, 'm = total number of data points'. Here, the statistical 'Plotting Position (YP)' would be equivalent to ARI, which is calculated for each of the flood events as per the following equation:

$$\text{ARI} = \frac{N + 0.2}{m - 0.4} \tag{3.8}$$

Where N is the number of years of the record, not the number of total data points (usually, for this method, the number of data points exceeds the number of years for record). Calculated ARI values are plotted against corresponding flood discharge values in a semi-log paper (discharge values in normal scale and ARI values in log scale). Finally, a best-fit curve (usually straight line) is drawn through the points and this best-fit curve is the relationship between ARIs and flood discharges, which can be later used for the extraction of ARI (for a particular flood magnitude) or flood magnitude for a particular ARI.

Worked Example 1

Calculate equivalent EY and ARI values for a flood having 1 in 2 years AEP.

Solution:

1 in 2 years AEP, is 50% (=1/2*100) AEP. That means AEP = 0.50. Using Equation 3.3:

$$EY = \log_e \{1 - AEP\ (\%)\} = -\log_e \{1 - 0.5\} = -\log (0.5)$$
$$= 0.693$$

So, ARI = 1/EY = 1/0.693 = 1.44 years

Worked Example 2

Calculate equivalent EY and AEP values for a flood having 1 in 5 years ARI.

Solution: EY = 1/ARI = 1/5 = 0.2. Now, using Equation 3.3:

$$EY = -\log_e \{1 - AEP\ (\%)\} = 0.2$$
$$\log_e (1 - AEP) = -0.20$$
$$(1 - AEP) = e^{-0.20} = 0.8187$$
$$AEP = 1 - 0.8187 = 0.1813.\ \text{So, AEP} = 18.13\%$$

Worked Example 3

From the recorded flood magnitudes presented in Tables 3.1 & 3.2, through annual series flood frequency analysis (graphical method), estimate the flood magnitudes for AEPs of 0.01 and 0.03.

Solution:

Annual maximum flood discharges are shown in the table below:

Year	Flood (m³/s)
2000	106
2001	110
2002	89
2003	86
2004	95

Here, $N = 5$, to work out AEP values using Equation 3.4, we need an individual 'm' value. Flood discharges are ranked from highest to lowest, with m=1 for highest discharge.

Rank, m	Flood (m³/s)	$AEP = \dfrac{m - 0.4}{N + 0.2}$
1	110	0.12
2	106	0.31
3	95	0.50
4	89	0.69
5	86	0.88

After plotting AEP versus flood discharges, the following graph is produced:

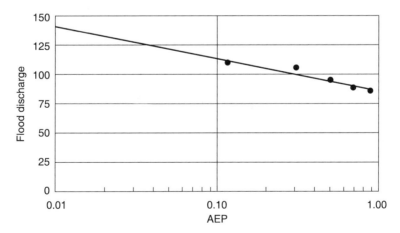

From the figure flood discharges for AEP=0.03 is 125 m³/s and for AEP=0.01 is 142 m³/s.

Worked Example 4

From the recorded flood magnitudes presented in Tables 3.1 and 3.2, through annual series flood frequency analysis (analytical method), estimate the flood magnitudes for AEPs of 0.01 and 0.05.

Solution:

Annual maximum flood values are as follow:

Year	Flood (m^3/s), X	$Log_{10}(X)$
2000	106	2.04
2001	110	2.03
2002	89	1.98
2003	86	1.95
2004	95	1.93

Using a scientific calculator or Excel functions, we can work out Mean (M), Standard Deviation (S) and Skew (g) values for the $Log_{10}(X)$ values. (Note: these values can be worked out using a scientific calculator or Excel functions: "AVERAGE" for mean, "STDEV" for standard deviation and "SKEW" for skewness).

For the given data, $M = 1.986$, $S = 0.047$ and $g = 0.22$

From the Table 3.4, for the given Skew (g) value,

K_Y (for 1% AEP) = 2.486 and K_Y (for 5% AEP) = 1.705 (Note: You need to do interpolation to get these values)

Now, using Equation 3.7, $\log Q_y = M + K_Y S$

$\log Q_{1\%} = M + K_{1\%}S = 1.986 + 2.486*0.047 = 2.103$

So, $Q_{1\%} = 10^{2.103} = 126.77 \ m^3/s$.

$\log Q_{5\%} = M + K_{5\%}S = 1.986 + 1.705*0.047 = 2.066$

So, $Q_{5\%} = 10^{2.066} = 116.41 \ m^3/s$.

Worked Example 5

An Annual Series flood frequency analysis has been undertaken for a stream flow recording station in Victoria. Analysis of the flow records has produced the following results: Mean (M) = 1.05, Standard Deviation (S) = 0.18, and Skewness (g) = –0.16. Calculate the runoff for the flood with an ARI = 500 years.

Solution:

From the Table 3.4, for the given Skew (g) value (–0.16), $K_{500 \ ARI} = 2.685$

(Note: You need to do interpolation between 2.637 and 2.757 to get this value).

Now, using Equation 3.7, $\log Q_y = M + K_Y S$

$$\log Q_{500} = M + K_{500} S = 1.05 + 2.685*0.18 = 1.533$$

So, $Q_{500} = 10^{1.533} = 34.14$ m³/s.

Worked Example 6

An Annual Series flood frequency analysis using analytical method has been undertaken for a stream flow recording station in Victoria. Analysis of the flow records has produced the following results: Mean (M) = 1.05, Standard Deviation (S) = 0.18, and 10 years ARI discharge (Q_{10}) = 18.37 m³/s. Determine the Skew (g) of the analysed data.

Solution:

From the given data, $M = 1.05$, $S = 0.18$ and $Q_{10} = 18.37$ m³/s.

Now, using Equation 3.7, $\log Q_y = M + K_y S$

$$\log (18.37) = M + K_{10} S = 1.05 + K_{10}*0.18 = 1.264$$

So, $K_{10} = 1.1895$

Now, from the Table 3.4, $g = -0.662$ (Note: You need to do interpolation between "−0.60" and "−0.70").

Worked Example 7

An Annual Series flood frequency analysis using analytical method has been undertaken for a stream flow recording station in Victoria. Analysis of the flow records has produced the following results: Mean (M) = 1.15, Standard Deviation (S) = 0.15, and discharge of AEP=0.99 is 7.5 m³/s. Determine the Skew (g) of the analysed data.

Solution:

From the given data, $M = 1.15$, $S = 0.15$ and $Q_{0.99} = 7.5$ m³/s.

Now, using Equation 3.7, $\log Q_y = M + K_y S$

$$\log (7.5) = M + K_{10} S\ 1.15 + K_{0.99}*0.15 = 0.875$$

So, $K_{0.99} = -1.83$

Now, from the Table 3.4, $g = 0.668$ (Note: You need to do interpolation between "0.60" and "0.70").

Worked Example 8

Recorded flood magnitudes for a stream in Victoria are presented below. Through 'Peak-over Threshold series' flood frequency analysis (use a flood threshold magnitude of 5.0 m³/s), estimate the flood magnitude with an ARI = 10 years.

Date	Flood (m³/s)
27/03/1965	5.1
27/03/1965	3.5
20/11/1964	11.0
16/03/1964	3.9
7/05/1963	4.5
31/10/1963	7.9
10/05/1962	17.0
15/06/1962	6.5
12/08/1962	3.0

Solution:

Considering all the flood values $\geq 5.0\,\text{m}^3/\text{s}$, there are five values (as shown in the following table). Floods are ranked with 'm=1' for the highest and 'm=5' for the lowest. Then corresponding ARI values are calculated using Equation 3.8 (shown in the last column of the following table).

Rank, m	Flood (m³/s)	$ARI = \dfrac{N + 0.2}{m - 0.4}$
1	17	7.00
2	11	2.63
3	7.9	1.62
4	6.5	1.17
5	5.1	0.91

Now these ARI and flood values are drawn in a semi-log paper (flood values in the normal axis and ARI values in the log axis). The following graph is produced.

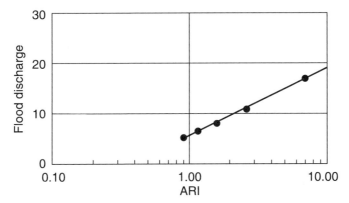

Now, drawing a best-fit line through the points and extending the line up to the ARI value "10", it is obtained that flood discharge for the 10 years ARI is 19.0 m³/s.

Worked Example 9

For the same data mentioned in the 'Worked example 8', estimate the flood magnitude with an ARI = 100 years.

Solution:

From the same graph of ARI versus flood discharge, extend the best-fit line up to the ARI value of "100" (Note: The graph axes need to be extended). The following graph is obtained and from the graph the 100 ARI flood is 32 m³/s.

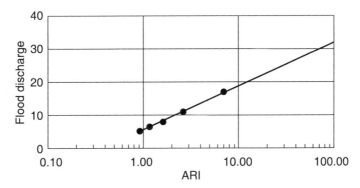

Reference

Ball, J., Babister, M., Nathan, R., Weeks, W., Weinmann, E., Retallick, M. and Testoni, I. (Editors) (2016). Australian Rainfall and Runoff: A Guide to Flood Estimation. © Commonwealth of Australia (Geoscience Australia).

Design Rainfall

4.1 Introduction

For any water resources related to infrastructure (i.e. pipe, irrigation channel, drainage canal, culvert, bridge, etc.) a prime design question is 'what should be the capacity of these structures?', in other words 'what should be the height/level/opening, width/diameter, slope and discharge capacity of these structures?'. These questions may arise even for some non-water related structures (i.e. building roof and its drainage). Also, in many countries, urban roads are designed to carry stormwater in addition to their main purpose of facilitating traffic movement. In general, there is no single answer for these sorts of capacity related questions. These answers depend on the level of acceptance of risk, societal value and financial affordability. In most cases, 'financial affordability' becomes the sole critical criteria in decision-making regarding these questions. Obviously, an ideal scenario would be that all these structures are capable of carrying any amount of rainwater (i.e. discharge/flood) without causing any nuisance, hindrance or adverse impacts to the users, neighbours or locality. However, such an ideal scenario is not achievable mainly due to enormous costs or, in other words, financial constraint. Accepting this reality of 'financial constraint', hydrologists and water engineers have been trying to minimise risks of flooding associated with these structures and to provide an informed solution to the users, community and authority. For example, for some constraint(s) a culvert might have been designed and built to carry a one-year maximum discharge only. In such case, it is the duty of the water engineers to inform the authority/community that this culvert is likely to be flooded once in a year.

To be able to evaluate a maximum discharge in a certain period (often years), the standard procedure is to perform statistical analysis with the recorded historical discharge data at the location of the point of interest.

However, to measure the discharge is always troublesome and expensive; as such, the discharge data is seldom available. Also, in many cases, the accuracy of the discharge measurements is questionable. On the other hand, rainfall measurements are easy. For these reasons, engineers traditionally depended on rainfall measurements and performed statistical analysis on historical recorded rainfall data. Through statistical analysis, different types (i.e. magnitudes and intensities) of rainfalls were derived. These derived rainfalls are called 'design rainfall' and widely used as standard rainfalls. Eventually, using these design rainfall values, discharges/floods of different standard magnitudes are calculated (procedures of which will be explained in a later chapter). These calculated discharges are called 'design discharge/flood'. However, in some cases where good quality of enough measured discharge data is available, statistical analysis is performed on measured discharge data and 'design discharge' is directly evaluated from the measured discharge data (as opposed to calculating discharge from derived rainfall values).

4.2 Intensity-Duration Relationship

In water resources applications, intensity of rainfall is a critical parameter influencing many design outcomes. However, evaluation of rainfall intensity requires critical thinking as it is significantly influenced by the considered period of rainfall. Consider a typical rainfall pattern (rainfall hyetograph), as shown in Figure 4.1. The rainfall event occurred for three hours and it was measured at 6 minutes intervals. For the task of deriving maximum rainfall intensity, if only 6 minutes' duration is considered then the maximum magnitude of rainfall in 6 minutes is 12.8 mm (during the early part of the event). This equates to an intensity of (=12.8/6*60) = 128 mm/hr (mm/hr is a standard unit of rainfall intensity, although some countries use cm/hr or inch/hr). However, if only 12 minutes' duration is considered, then the maximum rainfall in 12 minutes is 21.3 mm, which equates to an average intensity of (=21.3/12*60) = 106.5 mm/hr. Similarly, considering 30 minutes' maximum rainfall, which is 41.5 mm, the average intensity is (=41.5/30*60) = 83 mm/hr; considering 60 minutes' maximum rainfall, which is 66.5 mm, the average intensity is (=66.5/60*60) = 66.5 mm/hr; considering 90 minutes' maximum rainfall, which is 82.8 mm, the average intensity is (=82.8/90*60) = 55.2 mm/hr and considering 120 minutes' maximum rainfall, which is 89.6 mm, the average intensity is (=89.6/120*60) = 44.8 mm/hr. Table 4.1 shows the total rainfall depths and corresponding maximum rainfall intensities for different rainfall periods for the same rain event.

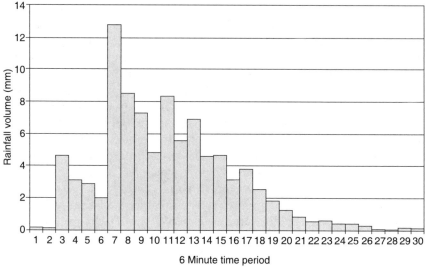

Figure 4.1. Rainfall hyetograph for a rain event of 180 minutes

Table 4.1. Rainfall depths and intensities for different rainfall durations

Duration (min.)	6	12	30	60	90	120
Rainfall depth (mm)	12.8	21.3	41.5	66.5	82.8	89.6
Intensity (mm/hr)	128	106.5	83.0	66.5	55.2	44.8

From the table, it is clear that the maximum rainfall intensity decreases with the increase of rainfall duration. Figure 4.2 is the graphical representation of the values shown in the table. From the figure it is clear that the rainfall intensity decreases exponentially with the increase of rainfall duration.

4.3 Derivation of Design Rainfall

Design rainfalls are the rainfall intensities to be considered for the purpose of design in a particular location. These rainfalls are derived from rigorous statistical analyses of historical recorded rainfall data for a particular area. Determination of design rainfall is an important part of the hydrological design procedure. It is used as an input to the design of a wide range of hydraulic structures to determine design sizes and risks of overtopping or failure. Design rainfalls are dependent on considered rainfall 'time period' and ARI/AEP to be considered for a particular structure/system.

The Weather/Meteorology department in an individual country is responsible for the derivations of such design rainfalls. Methodology of analysis may be different for different countries. Engineers do not

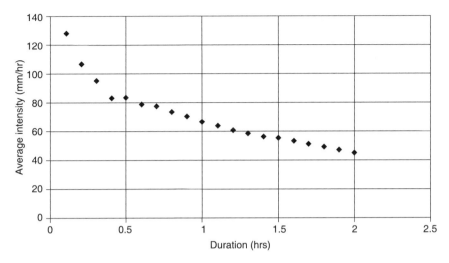

Figure 4.2. Variation of average rainfall intensity with the duration of rainfall

necessarily understand/know the details of these analyses. Standard design rainfalls are provided in the form of a chart and/or table.

Conducting frequency analysis as mentioned in Chapter 3, for each of the considered durations from several rainfall events, maximum rainfall intensities for different ARI/AEP events can be evaluated, and the shape of such intensities for a particular ARI/AEP is likely to be similar to Figure 4.2. However, for higher AEP/ARI events, intensities are likely to be higher and vice versa. For each standard ARI/AEP, there will be a separate curve similar to Figure 4.2. The set of relationship (between intensity and duration) curves for all the standard ARI/AEP events is called as IFD (Intensity-Frequency-Duration) or IDF (Intensity-Duration-Frequency) chart/table. Figure 4.3 is showing a typical IFD/IDF curve. If such curves are produced in log-log scale, often the curves become close to straight lines as shown in Figure 4.4. It is to be noted here that the standard ARI/AEP events are not universally the same, i.e. different countries use different standard measures. Precise derivation methods also vary from country to country.

In the following section, 'design rainfall' practices of some countries are outlined:

Australian Practice

Prior to late 2016, Australian Bureau of Meteorology used to provide IFD chart for any location (within Australia), which used to present rainfall intensities for different durations under 7 standard ARI categories (1 year, 2 year, 5 year, 10 year, 20 year, 50 year and 100 year), as shown in Figure 4.4. Since late 2016, Australian authorities have introduced new design

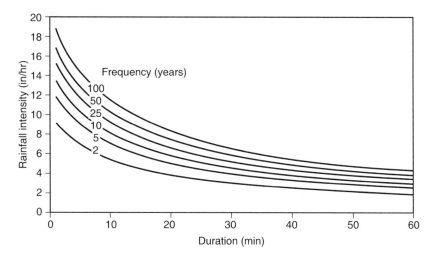

Figure 4.3. Typical set of IFD/IDF curve in normal scale

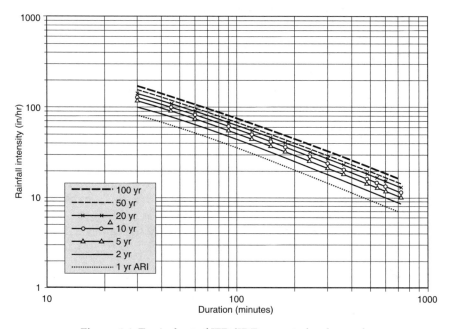

Figure 4.4. Typical set of IFD/IDF curve in log-log scale

guidelines (Australian Rainfall and Runoff 2016), which provide new IFD design rainfalls for any locality in Australia through the website: http://www.bom.gov.au/water/designRainfalls/revised-ifd/. By providing latitude and longitude of any locality, the online tool will provide IFD chart/table as rainfall intensities (or total depth) versus duration for different standard AEP/EY events under different categories ('Very Frequent', 'Frequent & Infrequent' and 'Rare'). There are options to view/export the design rainfalls in tabular or graphical format, as well as in the form of intensity (mm/hr) or depth (mm). Under 'very frequent' events the intensities/depths are presented for the following standard EY events: 12EY, 6EY, 3EY, 2EY, 1EY, 0.5EY and 0.2EY. Table 4.2 shows a typical 'very frequent' IFD table for an Australian locality (Latitude: 33.8625 S and Longitude: 151.2125 E) presented as rainfall depth (mm).

Table 4.2. Typical IFD table of rainfall depths (mm) for frequent events

	Exceedance per Year (EY)							
Duration	*12EY*	*6EY*	*4EY*	*3EY*	*2EY*	*1EY*	*0.5EY#*	*0.2EY**
1 min	1.09	1.23	1.48	1.66	1.93	2.41	3.01	3.73
2 min	1.93	2.16	2.57	2.87	3.29	4.02	4.99	6.11
3 min	2.63	2.96	3.54	3.96	4.54	5.57	6.92	8.49
4 min	3.23	3.65	4.38	4.91	5.66	6.97	8.67	10.7
5 min	3.74	4.24	5.12	5.76	6.66	8.23	10.3	12.7
10 min	5.59	6.4	7.83	8.86	10.3	13	16.2	20.2
15 min	6.82	7.83	9.62	10.9	12.8	16.2	20.3	25.2
30 min	9.12	10.5	12.9	14.8	17.4	22.2	27.7	34.4
1 hour	11.7	13.5	16.7	19	22.5	28.8	35.9	44.3
2 hour	14.9	17.1	21.1	24.1	28.6	36.7	45.6	56.1
3 hour	17.1	19.7	24.4	27.8	33	42.5	52.7	65
6 hour	21.8	25.2	31.4	36.1	42.9	55.5	68.9	85.5
12 hour	28	32.7	41.2	47.5	56.8	73.9	92.2	116

#0.5EY represents to 2 year ARI IFD, not the 50% AEP IFD
*0.2EY represents to 5 year ARI IFD, not the 20% AEP IFD

Under 'frequent & infrequent' events the intensities/depths are presented for the following standard AEP events: 63.2%, 50%, 20%, 10%, 5%, 2% and 1%. Figure 4.5 shows a typical IFD chart under the category of 'frequent & infrequent' AEP events for an Australian locality presented as intensity (mm/hr). Figure 4.6 shows the same IFD rainfall values for the same locality, however presented as total depth of rainfall (mm).

Under 'rare' events the intensities/depths are presented for the following standard AEP events: 1 in 100, 1 in 200, 1 in 500, 1 in 1000 and 1

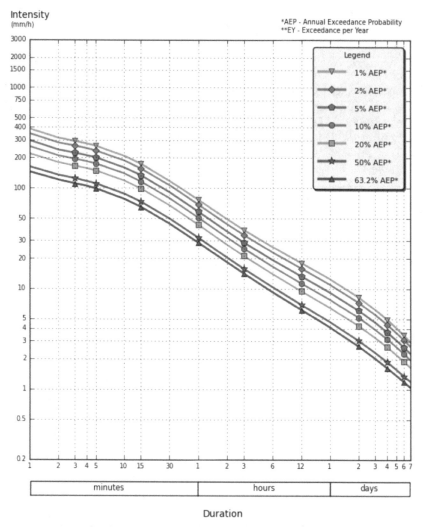

Figure 4.5. Typical Australian IFD chart for rainfall intensity

Color version at the end of the book

in 2000. Table 4.3 shows a typical 'rare' IFD table for the same Australian locality presented as rainfall intensity (mm/hr).

US Practice

US rainfall design charts and tables are provided by NOAA's Hydrometeorological Design Studies Center, through "Precipitation Frequency Data Server" available online at https://hdsc.nws.noaa.gov/

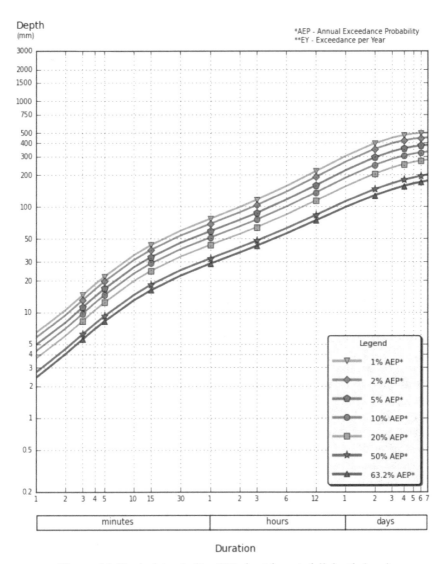

Figure 4.6. Typical Australian IFD chart for rainfall depth (mm)

Color version at the end of the book

hdsc/pfds/. In general, data provided by US authority is more versatile. They provide both AEP and ARI estimates, and values are provided with 90% confidence limits. Like Australian authority, there are options of selecting both rainfall depth and intensity; data can be extracted both in tabular and graphical formats. Also, users have the options of selecting a unit, either "English (inch)" or "Metric (mm)".

Table 4.3. Typical IFD table of rainfall intensity (mm/hr) for rare events

	Annual Exceedance Probability (1 in x)				
Duration	1 in 100	1 in 200	1 in 500	1 in 1000	1 in 2000
24 hour	12.5	13.5	15.3	16.7	18.1
48 hour	8.26	9.38	10.8	12	13.2
72 hour	6.21	6.97	7.96	8.73	9.53
96 hour	4.94	5.48	6.21	6.77	7.34
120 hour	4.07	4.49	5.07	5.51	5.95
144 hour	3.44	3.78	4.28	4.65	5.02
168 hour	2.96	3.26	3.71	4.04	4.37

From the mentioned website, users are required to select a state. Once a state is selected, the next window will provide names of all the rainfall stations within the selected state through a dropdown menu. Users can also select a particular location/station through providing 'latitude and longitude', address or selecting on the interactive map. If the 'rainfall depth' is selected to be presented, then the chart is called 'Depth-Duration-Frequency (DDF)' curves and if the 'rainfall intensity' is selected, then the chart is called 'Intensity-Duration-Frequency (IDF)' curves. If 'Annual Maximum Series' analysis is selected, then the result is presented in relation to AEPs: 1 in 2, 1 in 5, 1 in 10, 1 in 25, 1 in 50, 1 in 100, 1 in 200, 1 in 500 and 1 in 1000 years. If 'Partial Series (Peak-Over-Threshold)' analysis is selected, then the result is presented in relation to ARIs: 1 year, 2 year, 5 year, 10 year, 25 year, 50 year, 100 year, 200 year, 500 year and 1000 year. Graphical presentation includes both 'Duration values on x-axis while depth/intensity on y-axis' (Figure 4.7) and 'AEP/ARI values on x-axis while depth/intensity on y-axis' (Figure 4.8) formats. Figure 4.7 shows the DDF curves in relation to ARIs for a locality, where duration is shown on the x-axis and Figure 4.8 is for the same rainfall depths, except the ARI values are shown on the x-axis. In the tabular format, design rainfalls/depths are presented with a range of values, representing lower and upper bounds of the 90% confidence limit. In the graphical format, this range is shown with two lines, above and below the median line presenting depth/intensity for a particular time period, as shown in Figure 4.9.

Malaysian Practice

Malaysian government has published 'Urban Stormwater Management Manual for Malaysia, MSMA (2012)', which describes the procedure for deriving design rainfall intensities using an empirical equation. The proposed equation was fitted to the IDF data for 135 major cities in

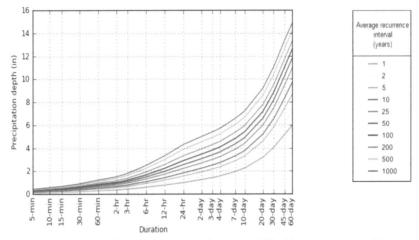

Figure 4.7. Typical DDF curves in relation to ARIs (duration on x-axis)

Color version at the end of the book

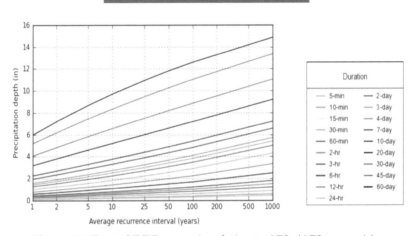

Figure 4.8. Typical DDF curves in relation to ARIs (ARI on x-axis)

Color version at the end of the book

Malaysia. This newly introduced equation is a modification of an earlier polynomial equation, which has been in use since 2000. The Equation is as follows:

$$i = \frac{\lambda T^{\kappa}}{(d + \theta)^{\eta}} \tag{4.1}$$

where i is the average rainfall intensity (mm/hr), T is the average return interval (years) and d is the storm duration (hours). λ, κ, η and θ are the fitting constants dependent on the city/location. The manual has provided

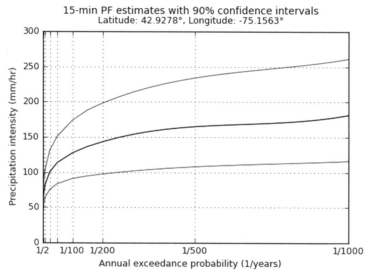

Figure 4.9. A typical graphical presentation of design rainfall intensity showing 90% confidence limit (upper and lower bounds) for a 15 minutes rainfall

values of these fitting constants for 135 locations within Malaysia. This equation is valid for the durations ranging from 5 minutes to 72 hours. Figure 4.10 shows a typical Malaysian IDF curve.

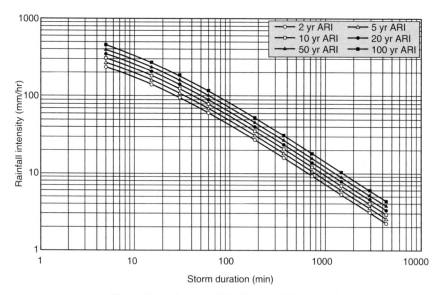

Figure 4.10. A typical Malaysian IDF curve

4.4 Temporal Pattern

IFD provides the design rainfall (i.e. average rainfall intensity) for different durations. For smaller duration (5~15 minutes) rainfalls, it is acceptable to consider a same intensity rainfall across the whole time period. However, for longer period rainfalls, it is unlikely that rainfall amounts/intensity will be equally distributed across the whole time period. For example, 1 hour 50 years design rainfall for the Brisbane city is 100 mm/hour. It does not mean that the design rainfall would be constant as 10 mm in each 6 minute interval (or 20 mm in each 12 minute interval) for 1 hour. In reality, that 100 mm/hr rainfall is an average amount of rainfall in a whole hour and during that period the rainfall pattern is likely to fluctuate. For example, in the following table, for a particular rainfall event, rainfall amounts in 5 minute intervals for 30 minutes are presented:

Time (minutes)	5	10	15	20	25	30
Rainfall (mm)	5	8	10	12	7	4
Rainfall Intensity (mm/hr)	60	96	120	144	84	48

Rainfall amounts increased to a peak (144 mm/hr) from a lower value (60 mm/hr) and then dropped down to 48 mm/hr. The total amount of rainfall in 30 minutes is 46 mm with an average intensity of 92 mm/hr.

Temporal pattern provides a likely distribution of design rainfall across the whole rainfall period. Temporal distribution has a significant impact in the calculations of runoff produced from a catchment for a particular design rainfall having higher duration. Through rigorous statistical analysis expected/design temporal distribution for a particular location can be established. For Malaysia, the 'Urban Stormwater Management Manual' has provided design temporal patterns for five different regions within Malaysia. This pattern reveals fractions of the total rainfall in different time intervals/blocks. Table 4.4 shows some portion of standard temporal patterns for the Malaysian capital, Kuala Lumpur, for different time periods.

Similarly, in Australia, in regard to standard temporal patterns, 8 major zones were established. Under each major zone there were several sub-zones, for each of which different temporal patterns were provided by the Australian Rainfall and Runoff, ARR 1987 (Pilgrim, 2001). However, after the introduction of the latest Australian Rainfall and Runoff, ARR (2016), those temporal patterns were made obsolete. Rather, the latest ARR suggests the use of 10 ensemble rainfall patterns for a particular locality from 'ARR Data Hub' provided online (http://data.arr-software.org/). Then it recommends to calculate runoffs using all these 10 ensemble rainfall patterns and eventually to take a mean of these 10 calculations.

Table 4.4. Temporal pattern for Kuala Lumpur up to 6 hours rainfall

No. of block	Storm duration (hr)				
	0.25	0.5	1	3	6
1	0.184	0.097	0.056	0.048	0.033
2	0.448	0.161	0.061	0.06	0.045
3	0.368	0.400	0.065	0.078	0.092
4		0.164	0.096	0.095	0.096
5		0.106	0.106	0.097	0.107
6		0.072	0.164	0.175	0.161
7			0.108	0.116	0.118
8			0.103	0.096	0.102
9			0.068	0.093	0.096
10			0.065	0.062	0.091
11			0.058	0.05	0.037
12			0.050	0.03	0.023

Worked Example 1

Figure 4.9 shows AEP versus rainfall intensity graphs (with 90% confidence limit) for a 15 minute rainfall for a location in USA. From the graph, find the expected ranges of 1 in 200 AEP and 1 in 2 AEP rainfalls for the same location. Comment on the answers.

Solution:

From the graph, for 1 in 200 AEP:

The lower limit is 100 mm/hr and the higher limit is 200 mm/hr.

So, the range is 100~200 mm/hr.

For 1 in 2 AEP:

The lower limit is 58 mm/hr and the higher limit is 90 mm/hr.

So, the range is 58~90 mm/hr

For 1 in 200 AEP, the range is much higher, as this is an extreme event, for which prediction will not be very accurate and is likely to have very wide range in comparison to a frequent event.

Worked Example 2

Calculate design rainfall intensities for the Malaysian city of Kuala Lumpur (use Ladang Edinburg station) for 1 year 15 minutes, 5 years 1 hour and 10 years 3 hours events. Distribute the design rainfalls as per the temporal pattern provided in Table 4.4.

Solution:

Malaysian 'Urban Stormwater Management Manual' provides different values of 'λ, κ, η and θ' for 0.5~12 months and 2~100 years ARI rainfalls. For 1 year 15 minutes rainfall, the values of the coefficients, λ, κ, η and θ are 64.50, 0.2751, 0.1814 and 0.8329, respectively (Table 2.B2 of MSMA, 2012).

Using Equation 4.1: $i = \dfrac{\lambda T^{\kappa}}{(d + \theta)^{\eta}}$, using above mentioned coefficients and

$T=1, d=0.25$, the 1 year 15 minutes rainfall intensity is 129.92 mm/hr.

For 5 year 1 hour and 10 year 3 hour rainfalls, the values of the coefficients, λ, κ, η and θ are 63.483, 0.146, 0.21, 0.83, respectively (Table 2.B1). Using these coefficients and Equation 4.1, considering $T=5$ and 10, $d= 1$ and 3,

5 year 1 hour rainfall intensity = 68.55 mm/hr and

10 year 3 hour rainfall intensity = 33.75 mm/hr.

Now, to find temporal distribution, we need to get total rainfall depths:

Rainfall depth for 1 year 15 minutes rainfall = 32.48 mm

Rainfall depth for 5 year 1 hour rainfall = 68.55 mm

Rainfall depth for 10 year 3 hour rainfall = 101.25 mm

Now, distributing these depths as per the distribution provided in Table 4.4.

Rainfall distribution for 1 year 15 minutes rainfall

No. of block	Time (minutes)	Fraction	Fraction*Total rainfall depth	Distributed rainfall depth (mm)
1	5	0.184	=0.184*32.48	5.98
2	10	0.448	=0.448*32.48	14.55
3	15	0.368	=0.368*32.48	11.95

Rainfall distribution for 5 year 1 hour rainfall

No. of block	Time (minutes)	Fraction	Fraction*Total rainfall depth	Distributed rainfall depth (mm)
1	5	0.056	=0.056*68.55	3.84
2	10	0.061	=0.061*68.55	4.18
3	15	0.065	=0.065*68.55	4.46
4	20	0.096	=0.096*68.55	6.58

(Contd.)

(Contd.)

No. of block	Time (minutes)	Fraction	Fraction*Total rainfall depth	Distributed rainfall depth (mm)
5	25	0.106	=0.106*68.55	7.27
6	30	0.164	=0.164*68.55	11.24
7	35	0.108	=0.108*68.55	7.40
8	40	0.103	=0.103*68.55	7.06
9	45	0.068	=0.068*68.55	4.66
10	50	0.065	=0.065*68.55	4.46
11	55	0.058	=0.058*68.55	3.98
12	60	0.050	=0.050*68.55	3.43

Rainfall distribution for 10 year 3 hour rainfall

No. of block	Time (minutes)	Fraction	Fraction*Total rainfall depth	Distributed rainfall depth (mm)
1	15	0.048	=0.048*101.25	4.86
2	30	0.06	=0.06*101.25	6.08
3	45	0.078	=0.078*101.25	7.90
4	60	0.095	=0.095*101.25	9.62
5	75	0.097	=0.097*101.25	9.82
6	90	0.175	=0.175*101.25	17.72
7	105	0.116	=0.116*101.25	11.75
8	120	0.096	=0.096*101.25	9.72
9	135	0.093	=0.093*101.25	9.42
10	150	0.062	=0.062*101.25	6.28
11	165	0.05	=0.05*101.25	5.06
12	180	0.03	=0.03*101.25	3.04

References

MSMA (2012). Urban Stormwater Management Manual for Malaysia, Department of Irrigation and Drainage, Government of Malaysia. https://www.water.gov.my/jps/resources/PDF/MSMA2ndEdition_august_2012.pdf.

Pilgrim, D.H. (2001). Australian Rainfall and Runoff: A Guide to Flood Estimation, Vol. 1. Institution of Engineers Australia, Barton, ACT. ISBN: 0858257440.

Deterministic Flow/Flood Estimations

5.1 Introduction

First of all, it is necessary to understand the difference between flow/discharge and flood. The flow/discharge is any amount of water discharge running through the stream/river; the water discharge might have been generated from the adjoining catchment or either coming from upstream and/or from adjoining groundwater. All discharges/flows may or may not produce flood; any flow which is overtopping the channel/river/stream bank(s) is termed as flood. In Chapter 3, we discussed 'probabilistic flood estimations', which are mainly applicable to floods (i.e. flow above certain magnitude). Probabilistic flood estimation mainly reveals the probability of a certain magnitude event happening; in other words, what is the expected time period for a certain magnitude flood to reoccur at a particular river/stream/channel location. Such probabilistic estimations do not reveal the exact amount of flow/discharge occurring under certain rainfall scenario, which is at times needed by stakeholders. For example, a certain amount of rainfall is expected to occur on a catchment and the authority wants to know how much flow will be generating at the catchment outlet due to this impending amount of rainfall. Probabilistic estimation is unable to provide such information. Due to advancements in different climate/rainfall models, including general circulation model (GCM), prediction of short-term (up to a week) rainfall has achieved a significant level of accuracy. As such, this has become a common practice to calculate expected runoff/flow from an impending rainfall amount for the purpose of forecasting floods. For this purpose, predicted rainfall amounts are extracted from other suitable computer models (i.e. GCM). Then, by using this rainfall amount and catchment properties, runoff from this rainfall amount is estimated using the deterministic method discussed in the following sections.

5.2 Hydrograph Details

A complete outcome of deterministic flood/flow estimation is a runoff hydrograph from a particular rain event. Figure 5.1 shows a typical catchment, including stream networks and its runoff/hydrograph generation process. The upper left corner of the figure shows the catchment (the thick outer line is the catchment boundary and thin inner lines are the stream network) on which the rainfall is occurring. Dashed lines within the catchment denote the sub-areas draining to the catchment outlet at each time step (Δt). The lower right corner of the figure shows generated runoff magnitudes for a particular rain event after different time steps and arrows from the dashed lines to the runoff hydrograph are the corresponding runoff at the catchment outlet after different time steps. Runoff reaches to a peak when raindrops from the farthest point of the catchment reach to the catchment outlet, while rainfall is continuing. If rainfall amount/intensity starts dropping after reaching the peak, then runoff will also drop after reaching the peak. However, if the same intensity rainfall (which caused the peak) continues for a longer period of time, the hydrograph peak will continue as flat at maximum magnitude for a duration equals to peak-causing rainfall continued until commencement of recession.

Figure 5.2 shows a typical hydrograph with associated components, as well as rainfall event. The solid conical shaped line is the runoff magnitudes at different times and the consecutive bars are the rainfall

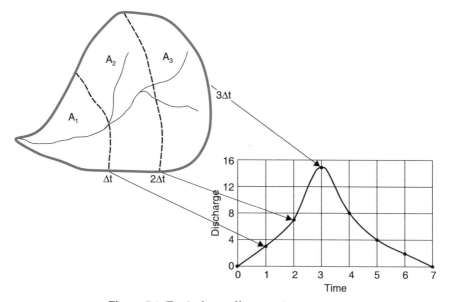

Figure 5.1. Typical runoff generation process

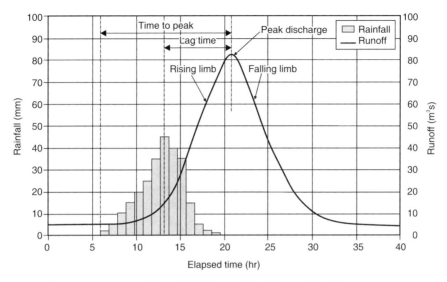

Figure 5.2. A typical runoff hydrograph with associated components

amounts. The units for time, rainfall and runoff are shown in hour (hr), millimetre (mm) and m³/s. However, these units can be expressed in any suitable unit. Most common units for time are minutes (min) and hour (hr); obviously 'min' for small catchments and 'hr' for large catchments. From the figure, it is seen that there was a small runoff amount, even before the rainfall started. This amount is referred to as 'base flow', which usually comes from groundwater storage. This amount is not necessarily to exist at all times and in urban settings in most cases this amount is zero. The period from the commencement of rain to the runoff peak is called 'time to peak' and the period between rainfall peak and runoff peak is called 'lag time'.

Factors Affecting Magnitude and Shape of the Runoff Hydrograph

Slope of the catchment: The steeper the catchment slope, the lower the rate of infiltration and faster the rate of runoff. So, for steeper catchment the runoff peak will be reached faster (i.e. shorter time to peak and lag time) with higher magnitude.

Soil type: Soils with large amounts of clay will absorb water slowly (i.e. less infiltration). On the other hand, soils having larger particle size (sand) will have larger infiltration capacities. Therefore, for clay dominated soil, the runoff peak will be reached slower (i.e. longer time to peak and lag time) with lower magnitude in comparison to sand dominated soil.

Soil condition: If the soil of the catchment surface is saturated, runoff peak will be more than in unsaturated (dry) soil. Also, runoff shape will be flatter.

Presence of vegetation: Catchment having dense vegetation canopies will intercept rainfall and will cause less runoff, as well as slower runoff peak (i.e. longer time to peak and lag time).

Land use: For catchment with impermeable surfaces (rock, pavement, roof in urban areas), runoff will occur quickly with higher magnitude than in a pervious surface.

Water use: Dams and reservoirs within catchment, slow down the rate of discharge at peak times as water is held back. Water extractions for industrial, irrigational and domestic uses cause a reduction in runoff discharge. Catchments having such infrastructure will have lower runoff, as well as slower runoff peak.

Drainage density: This is the length of river course per area of land within the catchment. The larger the amount of streams and rivers per area, the shorter distance water has to flow and the faster the rate of response will be. Catchments having higher drainage density, will have steeper hydrograph having higher peak.

Time of year/season: In summer, evapotranspiration rates are higher, reducing the amount of surface runoff.

5.3 Rational Method

Derivation of a complete runoff hydrograph for a catchment scale requires flow measurement at the catchment outlet, which is an expensive task. As such, hydrologists focused on finding a simpler way to get the peak runoff only. For most urban settings, if a smaller catchment is considered, the determination of peak runoff alone is good enough and is the most basic step for the deterministic flood/runoff estimation. As such, it has become a very common practice to calculate peak runoff/flood using the following simple formula known as 'Rational Formula':

$$Q = C * I * A \tag{5.1}$$

where Q is the runoff/discharge, C is the runoff coefficient (which depends on catchment properties), I is the rainfall intensity and A is the catchment area. This equation evolved from the basic rationale that volume of rainfall falling on a catchment during a specified time period is $I*A$, where I is the rainfall depth within the specified time period. The term C in equation 5.1, acts as a damper; i.e. $C=0$ when there is no runoff from a certain rain event and $C=1$ when all the rainfall turns into runoff from a certain rain event. Ideally, for an impervious surface (i.e. roof, road surface) C can be assumed as '1' (or close to 1) and runoff will happen as fast as rainfall. On the other hand, a small rain amount on a very porous surface having dry condition will be completely infiltrated into the soil, which represents a situation where C equals to zero. In reality, C=1 is very unusual, as even for an impervious surface it is likely to have other losses from rainfall such as depression and/or spilling. Also, C=0 is only possible for small amount of rain on a very pervious surface; for a moderately pervious (i.e. clayey) surface or for higher amounts of rainfall it is likely that there will be some runoff, which represents a non-zero C value. A proper determination of C values for pervious catchments in a real-life scenario has been a challenge and requires a vast knowledge/experience for a particular catchment. Even for the same surface C values will be different for different rainfall intensities; higher C values for higher rainfall intensities and lower C values for lower rainfall intensities, as in the case of higher/extreme rainfall intensities, the high burst(s) come after some prior small magnitude rainfalls, and by that time the catchment becomes almost saturated, contributing a higher ratio of runoff to rainfall. As such, in the case of design discharge determination, it has been a usual practice to assume a higher C value for the higher ARI/AEP events.

Equation 5.1 is often presented as per the following format in order to make it easy to memorise while having fixed units for the associated parameters:

$$Q = C * I * A / 360 \qquad (5.2)$$

where Q is in m³/s, I in mm/hr, A in hectare (Ha) and C is dimensionless. Nonetheless, any unit can be used and when using Equation 5.1 with proper conversions discharge/runoff can be estimated in any desired unit. This equation is often used for the purpose of estimation of design discharge for different water related infrastructures, and design rainfalls (as discussed in Chapter 4) are used for such estimations. As design rainfalls are dependent on two independent factors (duration and AEP/ARI), for a particular catchment design discharges will also depend on these factors, in addition to the factor related to catchment condition (i.e. runoff coefficient). Equation 5.2 can be represented as follows:

$$Q_y = C_y * I_y^d * A / 360 \qquad (5.3)$$

where y represents a particular AEP/ARI/EY event and d represents a particular duration. As mentioned earlier, C will depend on considered ARI/AEP event and I_y^d values are extracted from IFD/IDF chart for the selected locality. For the design purpose, a critical duration (when all of the catchment contributes to runoff at the catchment outlet) is selected as 'd'. Discharges/floods calculated through the above equation using design rainfalls become a probabilistic flood estimation, or pseudo-probabilistic estimation. Such design discharge estimations are widely used for the sizing design of various water-related structures (i.e. culvert, pipe, detention basin, channel, stormwater pit opening, weir, sluice gate, bridge opening, etc.). Determination of different 'C' values for different ARI/AEP events for different types of catchments requires extensive analysis, a task usually performed by the relevant government authority. After reviewing several relevant guidelines, Malaysian "Urban Stormwater Management Manual" recommended several C values for different land uses, as well as categories of rain event. Table 5.1 shows the list of recommended C values.

Table 5.1. Recommended values C for different surfaces

Land use	Runoff coefficient (C)	
	For minor event (<10 year ARI)	For major event (>10 year ARI)
Residential		
Bungalow	0.65	0.70
Semi-detached Bungalow	0.70	0.75
Terrace House	0.80	0.90
Flat and Apartment	0.80	0.85
Condominium	0.75	0.80
Commercial and Business Centres	0.90	0.95
Industrial	0.90	0.95
Sport Fields, Parks and Agriculture	0.30	0.40
Open Space		
Bare Soil (No Cover)	0.50	0.60
Grass Cover	0.40	0.50
Bush Cover	0.35	0.45
Forest Cover	0.30	0.40
Roads and Highways	0.95	0.95

Areal Reduction Factor

Usually, the measurement of rainfall intensity is a point estimation (i.e. measuring a rainfall intensity where the rain gauge is placed). However, it is likely that rainfall intensity will vary spatially, especially for a large catchment. To account for this spatial variation, the estimated runoff using the rational formula is usually multiplied by a factor called 'areal reduction factor (ARF)'. Different countries use different methods to derive ARF. For such ARF consideration, the Equation 5.3 becomes:

$$Q_y = ARF * C_y * I_y^d * A / 360 \tag{5.4}$$

5.4 Time of Concentration

Time of concentration (T_c) is defined as the travel time of rain drops from the most remote point in the catchment to the catchment outlet. These travel paths are usually not a straight line, and follow catchment topography and stream meandering (Figure 5.3). For the estimation of design discharge, the critical time for a catchment is its time of concentration and this is the 'time to peak' on a runoff hydrograph. There are several methods to calculate the time of concentration. It is to be noted here that all these formulas are empirical formulas, and, as such, should be used only with the specified units.

Pilgrim McDermott Formula

This is the simplest formula, which requires only the catchment area and, as such, is a very rough estimation and should be used only when other catchment related data is not available. This formula is suitable for a large homogeneous rural catchment having a moderate slope (McDermott and Pilgrim, 1982).

$$T_c = 0.76 * A^{0.38} \tag{5.5}$$

where T_c is the time of concentration (hrs) and A is the area of the catchment (km^2).

Bransby Williams Formula

This formula provides a more accurate estimation, however, it requires more data (i.e. stream length and catchment slope in addition to catchment area).

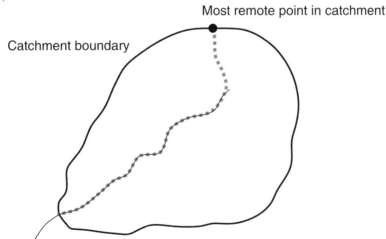

Figure 5.3. A typical catchment showing raindrop flow path to the outlet

$$T_c = \frac{FL}{A^{0.1}S_e^{0.2}} \tag{5.6}$$

where T_c is the time of concentration (mins), F is the unit conversion factor ($F = 58$ in SI method), L is the mainstream length (km), A is the catchment area (km^2) and S_e is the average slope of the catchment (m/km) which can be determined through the 'equal area slope' method. For the 'equal area slope' method, at first the surface elevations at different distances along the flow path are measured (often extracted from Digital Elevation Map, DEM) from the catchment outlet to the farthest point. Figure 5.4 shows a typical distance-elevation graph for a catchment, where the top solid line shows the elevation of channel bottom. Then, the area under the solid line is measured, often using the 'trapezoidal rule' discussed in Chapter 2. Finally a line is drawn from the catchment outlet (location '0' along the distance) to the farthest distance, forming a triangle underneath having a base equal to D and height equal to 'Equal Area Ordinate, EAO' (Figure 5.5) in a way that the area of this triangle is equal to the area (A) 'under the solid line' measured in Figure 5.4. The EAO can be calculated using Equation 5.7.

$$EAO = (A/D)*2 \tag{5.7}$$

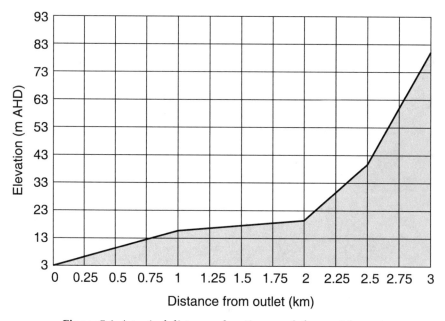

Figure 5.4. A typical distance-elevation graph for a catchment

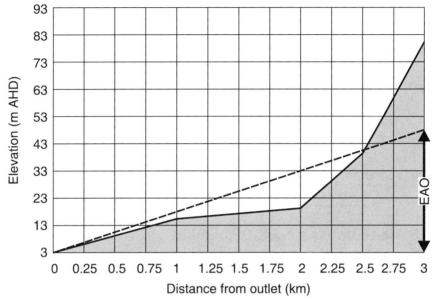

Figure 5.5. The determination of an Equal Area Ordinate, EAO

where *A* is the area of the surface under the elevation lines in Figure 5.4, *D* is the horizontal distance (in km) from the catchment outlet to the farthest point and EAO is the 'equal area ordinate' in metres. Then, the equal area slope S_e is equal to the EAO/D in m/km.

Kinematic Wave Equation

This is the most accurate equation among all the mentioned equations for the estimation of the time of concentration and, therefore, requires more data. It is suitable when the exact surface condition (i.e. cover) is known, in addition to the flow path length, slope and rainfall intensity. The equation is as follows:

$$T_c = \frac{6.94 * (L * n^*)^{0.6}}{I^{0.4} S^{0.3}} \tag{5.8}$$

where T_c is the time of concentration (mins), *L* is the flow path length (m), n^* is the surface roughness coefficient, *I* is the rainfall intensity (mm/hr) and *S* is the slope (m/m) of the flow path. Some recommended roughness coefficient values are provided in Table 5.2 (Pilgrim, 2001).

In reality, a catchment may comprise of different travel path systems, i.e. bare surface, impervious surface, gutter flow, piped flow, etc. The total time of concentration would be the sum of all the individual travel times.

Table 5.2. Recommended values of surface roughness coefficient

Surface type	Roughness coefficient
Concrete or Asphalt	0.010-0.013
Bare Sand	0.010-0.016
Gravel Surface	0.012-0.030
Bare Soil	0.012-0.033
Sparse Vegetation	0.053-0.130
Grass Prairie	0.100-0.200
Lawns	0.170-0.480

5.5 Non-homogeneous Catchment

In an urban setting, the catchments are usually comprised of different land uses, such as roof, road/footpath/driveway, grassd lawn and park, and these surfaces have different 'Runoff Coefficient (C)' values. As such, it is recommended that an equivalent or average C value should be determined for such catchment. Through measuring the area of each land use and assigning a corresponding C value, the following equation can be used to determine an average C value:

$$C_{avg} = \frac{\sum_{i=1}^{i=n} C_i A_i}{\sum_{i=1}^{i=n} A_i} \tag{5.9}$$

where C_i is the runoff coefficient for the ith catchment having an area of A_i.

5.6 Partial Area Effect

In some cases, two (or more) tributaries within a catchment drain from two (or more) different catchment types (i.e. land uses) and flow to a single outlet. In such a scenario, to determine a critical time period is not straight forward. This is because different sub-catchments have different travel times. For example, Figure 5.6 is showing a catchment with two tributaries (A & B) which are draining to a single outlet. Suppose, sub-catchment A is smaller, having a higher C value, on the other hand, sub-catchment B is larger, having a lower C value. As such, the time of concentration for the sub-catchment A (T_A) would be smaller than the time of concentration for the sub-catchment B (T_B). The time of concentration for the total catchment would also be, T_B. In this scenario, without doing a detailed calculation, it is not possible to know which time of concentration will

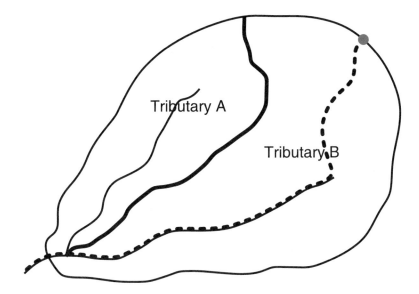

Figure 5.6. Typical catchment having two tributaries

provide a critical (i.e. worse) design discharge at the catchment outlet. At times, it is also possible that the consideration of T_A is providing a higher design discharge, due to the fact that the lower time period provides a higher design rainfall intensity from the IFD. In such cases, in order to determine the critical design discharge for a certain return period/AEP, it is necessary that the design discharges should be calculated considering both the time periods (i.e. T_A and T_B) separately and the one which produces a higher discharge is the design discharge for the catchment for that particular ARI/AEP. When the smaller time period (i.e. T_A in this case) is considered, the other sub-catchment (which takes a longer time to drain to the outlet) will not drain from its full area, rather, it will only drain from a partial area, depending on the ratio of T_A and T_B (Figure 5.7). Due to consideration of such partial area, this method is known as 'Partial Area Effect'.

The mathematical calculations for these two sub-catchments scenarios can be described as follows:

For two sub-catchments, there will be two sets of calculations; one for each time of concentration. Usually, all the associated C and area (A) values are given for such calculations, as such C_A, C_B, A_A and A_B values are known.

Case I: Consider T_A

Get corresponding design rainfall I_I from IFD table for the desired AEP/

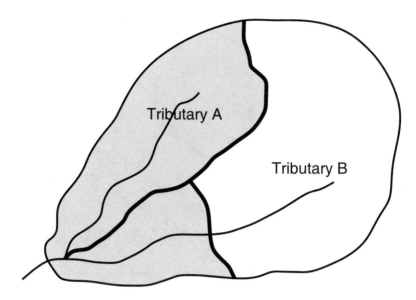

Figure 5.7. Typical example of partial draining from sub-catchment *B*

ARI. Using the rational formula, the total discharge at the catchment outlet would be:

$$Q_I = 1/360*[C_A*I_I*A_A + C_B*I_I*A_B*(T_A/T_B)]$$

here multiplication with (T_A/T_B) is for the consideration of the partial area effect.

Case II: Consider T_B

Get corresponding design rainfall I_{II} from IFD table for the same AEP/ARI. Using the rational formula, the total discharge at the catchment outlet would be:

$$Q_{II} = 1/360*[C_A*I_{II}*A_A + C_B*I_{II}*A_B]$$

The final design discharge for the desired AEP/ARI would be the highest among all the calculated Q values (in this case higher from Q_I and Q_{II}).

5.7 Composite Catchment

The above-mentioned partial area method might be good enough for catchments that have only two sub-catchments and with simple configurations, where such determination of partial area based on only linear ratio of time of concentrations is valid. However, most of the urban catchments are much more complex, comprising more than two sub-catchments and consideration of such partial area only based on linear ratio of travel times is likely to be erroneous. Accumulations of such errors will grow further if more sub-catchments are considered. Figure 5.8 shows a typical composite catchment having three sub-catchments (1, 2 and 3) with different characteristics and travel times to the final outlet, where internal solid lines showing creek/stream network and dashed lines are sub-catchment boundaries. For a relatively shorter rain event, the runoff peak from sub-catchment 3 will reach first as it is the nearest from the outlet, also its magnitude will be smallest (due to it being the smallest in size). Then, the runoff peak from sub-catchment 2 will arrive, being the second nearest from the outlet, and it will have a moderate magnitude. Runoff peak from sub-catchment 1 will arrive last (as it is farthest) with highest magnitude (due to its largest size). Typical runoff hydrographs from such catchment is shown in Figure 5.9.

As the rational formula provides only the peak value from a runoff hydrograph and, in this scenario, three runoff peaks are occurring in three different times, it will not be correct to add all the three peak values to get the total runoff. The 'partial area method' is an easy way of overcoming such an issue. It considers the peak from one sub-catchment, however adds only a portion of the peak runoff from the other catchment(s) through applying the linear proportion of the travel times. For the complex scenario mentioned-above, such a linear proportion will provide an erroneous result. As, for example, in this particular scenario, when the

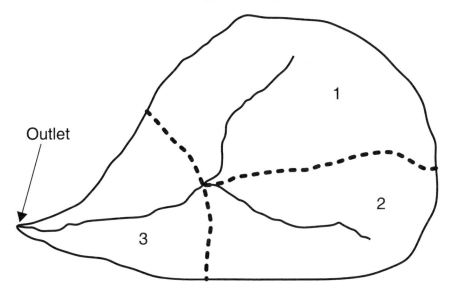

Figure 5.8. A typical composite catchment having three sub-catchments

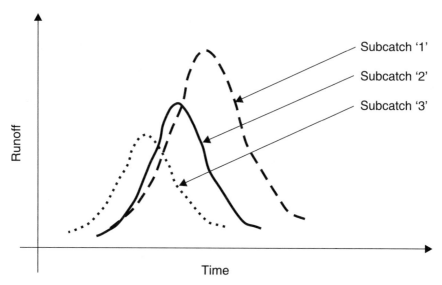

Figure 5.9. Hydrograph formation in a composite catchment

largest peak (from sub-catchment 1) arrived, by that time the contribution from sub-catchment 3 is almost zero (as rain had stopped much earlier). A more accurate way of calculating runoff from such catchment, is to get the coordinates of the complete hydrograph for each sub-catchment and add those coordinates as per their time of occurrences. The following section describes a widely-used method of deriving a complete runoff hydrograph, rather than getting only the peak value.

5.8 Unit Hydrograph Method

Theoretically, a unit hydrograph is a runoff response from a catchment to a unit input of rainfall excess (be noted that this is 'rainfall excess', not the 'total rainfall'). This allows for easy calculation of the response to any arbitrary rainfall, by simply multiplying the unit hydrograph ordinates with the arbitrary rainfall excess magnitude. Suppose a particular catchment from a particular rainfall of a certain duration having a rainfall excess of 1 unit, at the catchment outlet generated runoff values were measured, which is a complete hydrograph of 1 unit rainfall excess (unit hydrograph). Now for any amount of rainfall excess which occurs on the same catchment having the same duration, the total hydrograph can be easily derived by just multiplying the unit hydrograph ordinates with the new rainfall excess magnitude. In regard to the input unit rainfall and runoff values, 'cm' and 'm^3/s' are respectively used in SI system. However, in USA (English system) 'inch' and cfs (cubic feet per second) are used for the unit rainfall and runoff, respectively. In reality, for a particular catchment there should be many available unit hydrographs for different standard durations. For example, Figure 5.10 shows a unit hydrograph (lower dashed line) of 1 hour rainfall and runoff hydrograph (upper solid line) of a 1 hour 2 cm rainfall excess for the same catchment. It is to be noted that such multiplication is applicable only when the duration of the unit hydrograph input rain and the arbitrary rain are the same. In addition to performing simple multiplication, the available unit hydrograph(s) can be manipulated in order to derive the runoff hydrograph of any desired duration, provided that the available unit hydrograph duration is smaller and a regular fraction of arbitrary rainfall duration. For example, using the unit hydrograph shown in Figure 5.10, a total hydrograph can be derived for a 2 hour 2 cm rainfall excess (i.e. 1 cm per hour). In such case, the original unit hydrograph can be duplicated with a 1 hour shift and two hydrographs (dashed lines) of 1 cm each can be superimposed (i.e. added) in order to get a hydrograph of 2 cm 2 hour rainfall excess (solid line), as shown in Figure 5.11.

Similarly, if a unit hydrograph of 0.5 hour rainfall excess is available for a catchment, for a rainfall excess of 6 cm in 1. 5 hours (@ 2 cm per 0.5 hour), it can be derived as: First multiply the unit hydrograph ordinates by 2 (as the rate is 2 cm per 0.5 hour), then duplicate the same multiplied hydrograph twice (each shifted by 0.5 hour) replicating three hydrographs each for 2 cm 0.5 hour rainfall excess. Then, superimpose (i.e. add) all the three hydrographs in order to get the total hydrograph of 6 cm 1.5 hours rainfall excess. As mentioned earlier, from such unit hydrograph (1 cm 0.5 hour rainfall excess), it is not possible to derive a hydrograph of 0.25 hour or 0.75 hour.

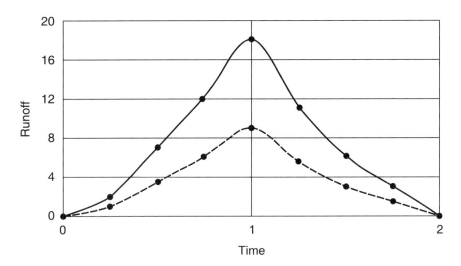

Figure 5.10. Typical example of the multiplication of unit hydrographs

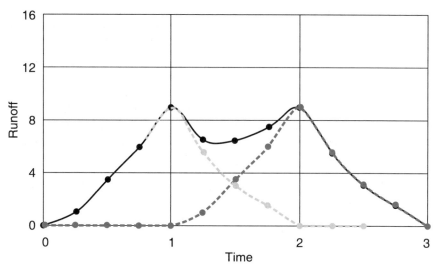

Figure 5.11. Typical example of superimposing unit hydrographs

5.9 Flood Modelling

Flood modelling has two phases: Hydrologic modelling and hydraulic modelling. Hydrologic modelling is the quantification of runoff/discharge from a catchment through providing catchment properties and rainfall data. Hydraulic modelling is the calculation of flood depth along a creek/river due to the runoff/discharge draining from the upstream catchment. In current days, flood modelling is an integral part of urban design, since with quantifications of expected flood discharges authorities can properly design urban infrastructures in order to keep residents safe from frequent flooding and associated nuisances. Therefore, only through such flood modelling can a flood warning system be developed. As mentioned earlier, through the use of complex GCM, rainfall up to one week can be accurately predicted. Through the use of such predicted rainfall for a particular catchment, applying flood modelling discharges at a particular catchment outlet and flood heights can also be predicted. This has enormous benefits for community safety, as in such case after receiving a warning of impending flood, the community/authorities/residents can make proper arrangements of evacuation, as well as prior relocation of their valuables. The hydrologic modelling and hydraulic modelling are discussed in detail in Chapter 8.

5.10   Time-Area Method

Time-area method is a widely-used method for hydrologic modelling. Here, it is explained in simple terms. In reality, different modelling tools tried to incorporate more complex parameters in order to enhance the accuracy of the calculations. At first, a catchment is sub-divided based on its time of concentration. To make it simple, the time step of this subdivision is kept the same as the time step (Δt) of rainfall. Then, from the rainfall amount in each time step (Δt), after deducting losses, rainfall excess (RE) amounts are calculated. Figure 5.12 shows a typical catchment subdivision. Suppose, for this catchment, that the time of concentration is 30 minutes and that it is divided into three sub-areas (A_1, A_2 and A_3) having Δt=10 minutes (note: It can be subdivided into any number, the more the better. However, as the number increases, computation time will also increase). Now, rainfall data needs to be collected/rearranged for every 10 minutes (i.e. Δt). Assume that there is a rainfall event for 40 minutes, which can be split into four different spells (each for 10 minutes) producing four different 'RE' amounts, as shown in Figure 5.13.

After the first time step (Δt), RE_1 from A_1 will only contribute at the catchment outlet. RE_1 from A_2 will only contribute after '$2\Delta t$' and RE_1 from A_3 will only contribute after '$3\Delta t$'. Similarly, RE_2 from A_1 will only contribute after '$2\Delta t$' and RE_2 from A_2 will only contribute after '$3\Delta t$'. Following the same step, contributing runoff from each sub-area after each time step can be calculated. Table 5.3 shows the runoff values from each sub-area at different time steps for the whole rainfall event having four spells of each 10 minutes. Now, the total runoff would be the sum of all the individual runoffs at a time step (i.e. added row wise).

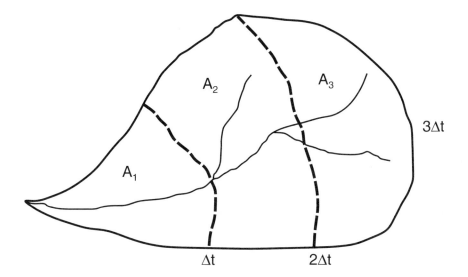

Figure 5.12. Catchment subdivision based on time step (Δt)

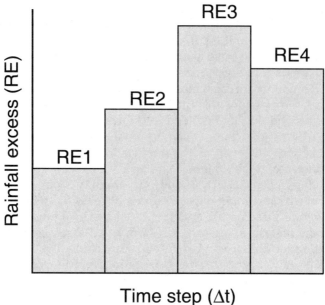

Figure 5.13. Rainfall excesses at different time steps

Table 5.3. Runoff calculations from individual sub-areas using time-area method

Time	Discharge	Runoff contribution from		
		A_1	A_2	A_3
Δt	Q_1	A_1*RE_1	0	0
$2\Delta t$	Q_2	A_1*RE_2	A_2*RE_1	0
$3\Delta t$	Q_3	A_1*RE_3	A_2*RE_2	A_3*RE_1
$4\Delta t$	Q_4	A_1*RE_4	A_2*RE_3	A_3*RE_2
$5\Delta t$	Q_5	0	A_2*RE_4	A_3*RE_3
$6\Delta t$	Q_6	0	0	A_3*RE_4
$7\Delta t$	Q_7	0	0	0

Finally, it yields to the following values:

$$Q_1 = A_1*RE_1$$

$$Q_2 = A_1*RE_2 + A_2*RE_1$$

$$Q_3 = A_1*RE_3 + A_2*RE_2 + A_3*RE_1$$

$$Q_4 = A_1*RE_4 + A_2*RE_3 + A_3*RE_2$$

$$Q_5 = A_2*\text{RE}_4 + A_3*\text{RE}_3$$

$$Q_6 = A_3*\text{RE}_4$$

5.11 Modelling Tools

There are several customised modelling tools available for hydrologic modelling. A few of those are:
1. RORBWin (http://www.harconsulting.com.au/software/rorb/)
2. WBNM (wbnm.com.aru)
3. XPRAFTS (http://www.innovyze.com/products/xprafts/)

Worked Example 1

A catchment of 1 ha has got 70% of its area covered by industrial and commercial buildings. The remaining 30% is open space having grass cover. Determine the average C value for the catchment for 5 year ARI and 20 year ARI events.

Solution:

Industrial and commercial area, $A_{ic} = 0.70$ ha
Open space area, $A_{os} = 0.30$ ha
For 5 year ARI: $C_{ic} = 0.90$, $C_{os} = 0.40$, and
For 20 year ARI: $C_{ic} = 0.95$, $C_{os} = 0.50$
Using Equation 5.9, average C value for 5 year ARI =

$$\frac{C_{ic} * A_{ic} + C_{os} * A_{os}}{A_{ic} + A_{os}} = \frac{0.9 * 0.7 + 0.4 * 0.3}{0.7 + 0.3} = 0.75$$

Again, average C value for 20 year ARI =

$$\frac{C_{ic} * A_{ic} + C_{os} * A_{os}}{A_{ic} + A_{os}} = \frac{0.95 * 0.7 + 0.5 * 0.3}{0.7 + 0.3} = 0.815$$

Worked Example 2

A catchment has an area of 4.5 km^2 and a mainstream flow path length of 3.0 km. Distance (from catchment outlet)-elevation profile along the flow path is shown in Figure 5.4 and the corresponding data is given in the table below. Calculate the equal area slope for this catchment.

Distance (km)	0	0.5	1.0	1.5	2.0	2.5	3.0
Elevation (m)	3	10	16	18	20	40	80

Solution:

First, get the height of each point from the lowest elevation (i.e. at the catchment outlet). From the above table, in the 'elevation' row, deducting '3' from each elevation will provide the height of each point from the lowest elevation. The following table shows the heights of all the points from the lowest elevation.

Distance (km)	0	0.5	1.0	1.5	2.0	2.5	3.0
Height from the lowest point (m)	0	7	13	15	17	37	77

Using the Trapezoidal rule, the area of the curve can be calculated as:

$$\text{Area} = \frac{0 + 77 + 2*(7 + 13 + 15 + 17 + 37)}{2}*0.5*1000 \text{ m}^2 = 63{,}750 \text{ m}^2$$

So, Equal Area Ordinate (EAO) = Area/(Flow Path Length)*2 = 63750/(3*1000)*2 = 42.50 m

So, Equal Area Slope, S_e = EAO/(Flow Path Length) = 42.5/3 = 14.17 m/km.

Worked Example 3

For the above-mentioned catchment, calculate the time of concentration (critical time) using both the Pilgrim and Bransby formulas.

Solution:

Using, Pilgrim formula, $T_c = 0.76 * A^{0.38}$

$$T_c = 0.76 * 4.5^{0.38} = 1.35 \text{ hrs} = 80.76 \text{ minutes}$$

Using, Bransby formula, $T_c = \dfrac{58S*L}{A^{0.1}S_e^{0.2}} = \dfrac{58*3}{4.5^{0.1}*14.17^{0.2}} = 88.10$ minutes

Worked Example 4

For the same catchment mentioned above, suppose the surface characteristics are known for having predominantly sparse vegetation (consider a middle value from the provided range), calculate the time of concentration using the 'Kinematic Wave Equation' for rainfall intensities of 50 mm/hr and 100 mm/hr.

Solution:

Unlike the previous two methods, in the 'Kinematic Wave Equation' the L is in meter and the slope (S) is in m/m. So, L=3000 m, S=0.0142 m/m.

From the Table 5.2, taking the middle value of the surface roughness range, n^*=0.0915.

Now, for rainfall intensity, I=50 mm/hr,

$$T_c = \frac{6.94 * (3000 * 0.0915)^{0.6}}{50^{0.4} 0.0142^{0.3}} = 151.1 \text{ minutes}$$

Again, for rainfall intensity, I=100 mm/hr,

$$T_c = \frac{6.94 * (3000 * 0.0915)^{0.6}}{100^{0.4} 0.0142^{0.3}} = 114.5 \text{ minutes}$$

Worked Example 5

For a catchment having an area of 1.0 km², a flow path length of 500 m, a slope 1 in 100 and a surface roughness (n^*) which equals to 0.012, a flood analysis should be conducted, for the design rainfall for 5% and 10% AEP events. To find the design rainfall, the 'critical time' of the catchment is required. Authority recommends to use the 'Kinematic Wave Equation' in order to calculate the 'critical time'. Determine the design rainfalls for the mentioned events from the partial IFD table for the catchment locality provided below.

Duration (mins)	Annual Exceedance Probability (AEP)				
	50% (mm/hr)	20% (mm/hr)	10% (mm/hr)	5% (mm/hr)	2% (mm/hr)
10	47.6	65	77	93	116
11	45.6	62	74	89	111
12	43.9	60	71	85	106
13	42.3	58	69	82	102
14	40.8	56	66	79	98
15	39.5	54	63	76	95
16	38.3	52	61	74	92
17	37.1	50	60	72	89
18	36.1	49.0	58	69	86
19	35.1	47.6	56	67	84
20	34.2	46.4	55	66	81

Solution:

Given values are:

L=500, S=1/100=0.01, and n^*= 0.012. As both the T (critical time) and

I (rainfall intensity) are unknown, an indirect method is necessary. In the kinematic wave equation, if all the unknown parameters are taken in the left hand side, the known parameters yield:

$$T * I^{0.4} = \frac{6.94 * (500 * 0.01)^{0.6}}{0.01^{0.3}} = 80.96$$

Now, in the given IFD table, all the I values are converted to $I^{0.4}$ and multiplied with the corresponding T value, which yields a table of T verses $T * I^{0.4}$' for different AEP events as shown below:

T (mins)	Annual Exceedance Probability (AEP)				
	50% $T*I^{0.4}$	20% $T*I^{0.4}$	10% $T*I^{0.4}$	5% $T*I^{0.4}$	2% $T*I^{0.4}$
10	70.3	79.7	85.2	91.9	100.4
11	69.1	78.2	83.9	90.3	98.7
12	68.1	77.2	82.5	88.7	96.9
13	67.1	76.1	81.6	87.4	95.4
14	66.1	75.1	80.2	86.1	93.9
15	65.3	74.0	78.7	84.8	92.7
16	64.5	72.9	77.7	83.9	91.5
17	63.7	71.7	77.2	83.0	90.3
18	63.0	71.1	76.1	81.6	89.1
19	62.3	70.3	75.1	80.6	88.3
20	61.6	69.6	74.5	80.2	87.0

Now, $T*I^{0.4}$ for this example is 80.96. Looking through 5% and 10% AEP columns; For 5% AEP, 80.96 corresponds to a T of between '18' and '19' and can be approximated as '18.5' minutes. Then for 10% AEP, 80.96 corresponds to a T of between '13' and '14' and can be approximated as close to '13.5'. Therefore, the times of concentration for 5% and 10% AEPs are 18.5 minutes and 13.5 minutes, respectively.

Now, from the original IFD table, the design rainfall for 5% AEP (i.e. for a T of 18.5 minutes) = 68 mm/hr (i.e. an average of '67' and '69'). Again, the design rainfall for 10% AEP (i.e. for a T of 13.5 minutes) = 67.5 mm/hr (an average of '66' and '69').

Worked Example 6

For a catchment having an area of 2.0 km², a 10% AEP design flood was estimated as 22.92 m³/s. For the catchment, an average C value was provided by the local council as '0.75'. Rainfall IFD table for the locality is provided in the previous example. Determine the time of concentration for the catchment.

Solution:

Given, area $(A) = 2$ km$^2 = 200$ ha, $C = 0.75$ and $Q = 22.92$ m^3/s. Using Equation 5.2, $Q = C * I * A/360$, $I = Q * 360/(C * A) = 22.92*360/(0.75*200) = 55$ mm/hr. Now, from the IFD table for the 10% AEP, corresponding time period for a rainfall intensity of 55 mm/hr is 20 minutes. So, the time of concentration for the catchment is 20 minutes.

Worked Example 7

Two tributaries are draining from two adjacent sub-catchments X and Y to a single outlet. Sub-catchment X is predominantly urban, having an area of 2 km^2 and an impervious area of 75%. On the other hand, sub-catchment Y is predominantly rural, having an area of 4 km^2 with an impervious area of 25%. Travel times for sub-catchment X and Y are 30 mins and 60 mins, respectively. For a 2% AEP flood, the local council recommends $C_{imp} = 0.9$ and $C_{per} = 0.6$. Determine the peak runoff from the combined catchment for a 2% AEP event. Partial IFD table for the locality is given below:

Duration (mins)	Annual Exceedance Probability (AEP)		
	5% (mm/hr)	2% (mm/hr)	1% (mm/hr)
10	93	116	136
15	76	95	110
30	52	64	75
32	50	62	72
34	48.4	60	69
36	46.8	58	67
38	45.3	56	65
40	43.9	54	62
45	40.8	50	58
50	38.2	47.0	54
55	36.0	44.2	51
60	34.0	41.8	48.1

Solution:

Sub-catchment areas in ha are, 200 ha and 400 ha. Average CA for the sub-catchment X,

$$(CA)_X = (0.9*0.75*200 + 0.6*0.25*200) = 165$$

Similarly, average CA for the sub-catchment Y,

$$(CA)_Y = (0.9*0.25*400 + 0.6*0.75*400) = 270$$

Case I: Consider T = 30 min

From the IFD table, for 2% AEP, design rainfall I_I =64 mm/hr.
So, runoff for Case I, Q_I

$$Q_I = [(CA)_X + (CA)_Y*(T_A/T_B)]*I_I/360$$
$$=[165+270*30/60]*64/360 = 53.33 \text{ m}^3/\text{s}$$

Case II: Consider T = 60 min

From the IFD table, for 2% AEP, design rainfall I_{II} =41.8 mm/hr.
So, runoff for Case II, Q_{II}

$$Q_{II} = [(CA)_X + (CA)_Y]*I_{II}/360 = [165+270]*41.8/360 = 50.51 \text{ m}^3/\text{s}$$

So, peak runoff for the catchment is 53.33 m^3/s.

Worked Example 8

An urban catchment has a total area of 320 ha, and is composed of three main tributaries and sub-catchments (X, Y, Z). The properties of the three sub-catchments are provided below. Partial IFD data for the locality is provided in the previous example. Determine the peak runoff from the catchment for a 2% AEP event. Impervious and pervious C values for the 2% AEP event are provided below.

Catchment	T_c (min.)	Total area (ha)	Impervious portion (%)	C_{perv}	C_{imperv}
X	30	160	60	0.60	0.95
Y	10	60	90	0.65	0.95
Z	15	100	70	0.70	0.95

Solution:

Average CA for the sub-catchment X,

$$(CA)_X = (0.95*0.6*160 + 0.6*0.4*160) = 129.6$$

Similarly, average CA for the sub-catchment Y,

$$(CA)_Y = (0.95*0.9*60 + 0.65*0.1*60) = 55.2$$

And, average CA for the sub-catchment Z,

$$(CA)_Z = (0.95*0.7*100 + 0.7*0.3*100) = 87.5$$

Case I: Consider $T = 30$ min

From the IFD table, for 2% AEP, design rainfall I_I=64 mm/hr.
Therefore, runoff for Case I, Q_I

$$Q_I = [(CA)_X + (CA)_Y + (CA)_Z]*I_I/360 = [129.6+55.2+87.5]*64/360$$

$$= 48.41 \text{ m}^3/\text{s}$$

Case II: Consider $T = 15$ min

From the IFD table, for 2% AEP, design rainfall I_{II} =95 mm/hr.
Therefore, runoff for Case II, Q_{II}

$$Q_{II} = [15/30*(CA)_X + (CA)_Y + (CA)_Z]*I_{II}/360$$

$$= [0.5*129.6 + 55.2 + 87.5]*95/360 = 54.76 \text{ m}^3/\text{s}$$

Case III: Consider $T = 10$ min

From the IFD table, for 2% AEP, design rainfall I_{II} =116 mm/hr.
Therefore, runoff for Case III, Q_{III}

$$Q_{III} = [10/30*(CA)_X + (CA)_Y + 10/15*(CA)_Z]*I_{III}/360$$

$$= [0.33*129.6 + 55.2 + 0.67*87.5]*116/360 = 50.50 \text{ m}^3/\text{s}$$

Therefore, peak runoff for the catchment is 54.76 m^3/s.

Worked Example 9

Table A shows the ordinates of a unit hydrograph (runoff generated from 1 cm rainfall excess of 1 hr) for a particular catchment. Rainfall and losses data for a particular rain event on the same catchment are shown in Table B and Table C, respectively. Calculate the ordinates of the total runoff hydrograph from the catchment for the mentioned rainfall event.

Table A

Time (hr)	Runoff (m³/s)
0	0
1	1
2	2
3	4
4	8
5	6
6	3
7	2
8	0

Table B

Time (hr)	Rainfall (cm)
0~1	2.0
1~2	3.5
2~3	2.0
3~4	1.0

Table C

Time (hr)	Loss (cm)
0~1	2.0
1~2	1.0
2~3	1.0
3~4	0.5

Solution:

First of all, the rainfall excess (RE) data needs to be evaluated from the given rainfall and losses data (Table B and C). The following table shows the RE data:

Time (hr)	Rainfall (cm)	Loss (cm)	RE (cm)
0~1	2	2	0
1~2	3.5	1	2.5
2~3	2	1	1
3~4	1	0.5	0.5

Now, the unit hydrograph data (Table A) needs to be multiplied with the respective RE values with proper lags (each by 1 hour). The final runoff hydrograph is the sum of all the 4 hydrographs from the four rainfall excesses, as shown below:

Time (hr)	UH*RE1	UH*RE2	UH*RE3	UH*RE4	Total
0	0	0	0	0	0
1	0	0	0	0	0
2	0	2.5	0	0	2.5
3	0	5	1	0	6
4	0	10	2	0.5	12.5
5	0	20	4	1	25
6	0	15	8	2	25
7	0	7.5	6	4	17.5
8	0	5	3	3	11
9	0	0	2	1.5	3.5
10	0	0	0	1	1
11	0	0	0	0	0

In the above table, the total hydrograph values are in m^3/sec.

Worked Example 10

Table D shows the ordinates of a unit hydrograph (runoff generated from 1 cm rainfall excess of 1 hr) for a particular catchment. Assume that the catchment was fully impervious and there was no infiltration loss; however, there is an initial loss of 1 cm during the first hour due to interception and depression storage losses. A rainfall occurred on the same catchment for a period of three hours having a total rainfall amount of 6 cm. It can be assumed that the rainfall was uniformly distributed over time. Calculate the ordinates of the total runoff hydrograph from the catchment for the mentioned rainfall event.

Table D

Time (hr)	Runoff (m^3s)
0	0
1	1
2	2
3	4
4	8
5	6
6	4
7	2
8	0

Solution:

As the total rainfall is distributed uniformly over time, it can be considered that the rainfall amount per hour was 2 cm. Evaluation of rainfall excess (RE) values are shown in the following table:

Time (hr)	Rainfall (cm)	Losses (cm)	RE (cm)
0~1	2	1	1
1~2	2	0	2
2~3	2	0	2

Now, the unit hydrograph data (Table D) needs to be multiplied with the respective RE values with proper lags (each by 1 hour). The final runoff hydrograph is the sum of all the 3 hydrographs from the three rainfall excesses, as shown below:

Time (hr)	UH*RE1	UH*RE2	UH*RE3	Total
0	0	0	0	0
1	1	0	0	1
2	2	2	0	4
3	4	4	2	10
4	8	8	4	20
5	6	16	8	30
6	4	12	16	32
7	2	8	12	22
8	0	4	8	12
9	0	0	4	4
10	0	0	0	0
11	0	0	0	0

In the above table, the total hydrograph values are in m^3/sec.

Worked Example 11

Table E shows the ordinates of two unit hydrographs; 'Unit hydrograph 1' is the runoff generated from 1 cm rainfall excess of 1 hr rainfall and 'Unit hydrograph 2' is the runoff generated from 1 cm rainfall excess of 0.5 hr rainfall for a particular catchment. A rainfall occurred on the same catchment for a period of 2.5 hours, having a total rainfall amount of 5 cm. It can be assumed that the rainfall was uniformly distributed over time. Assume that the catchment was fully impervious and there was no infiltration loss; however, there is an initial loss of 1 cm during the first hour due to interception and depression storage losses. Calculate the ordinates of the total runoff hydrograph from the catchment for the mentioned rainfall event.

Table E

Time (hr)	Runoff (m^3s)	
	Unit hydrograph 1	Unit hydrograph 2
0	0	0
0.5	0.5	1
1.0	1	2
1.5	4	8
2.0	4	3
2.5	3	1.5
3.0	2	0
3.5	1	0
4.0	0	0

Solution:

As the total rainfall is distributed uniformly over time, it can be considered that the rainfall amount per hour was 2 cm. Evaluation of rainfall excess (RE) values are shown in the following table:

Time (hr)	Rainfall (cm)	Losses (cm)	RE (cm)
0~1	2	1	1
1~2	2	0	2
2~2.5	1	0	1

For the total hydrograph, it can be considered that there were two rainfalls, each having 1 hr duration, followed by another rainfall of 0.5 hr. So, three hydrographs can be added as (1 hr + 1 hr + 0.5 hr) with proper lags, as shown below:

Time (hr)	UH1*RE1	UH1*RE2	UH2*RE3	Total
0	0	0	0	0
0.5	0.5	0	0	0.5
1	1	0	0	1
1.5	4	1	0	5
2	4	2	0	6
2.5	3	8	1	12
3	2	8	2	12
3.5	1	6	8	15
4	0	4	3	7
4.5	0	2	1.5	3.5
5	0	0	0	0

In the above table, the total hydrograph values are in m^3/sec.

Worked Example 12

A catchment area (shown below) has a total travel time of 6 mins. The catchment is subdivided as per equal time area, having $\Delta t=2$ mins. Areas are given as below:

A_1 (ha)	A_2 (ha)	A_3 (ha)
0.1	0.25	0.6

A rainfall occurred on this catchment with the following 'Rainfall Excesses':

Time (min.)	2	4	6
Rainfall Excess (mm)	1	3	2

Calculate the runoff generated at the outlet of this catchment using the 'Time-Area Method'.

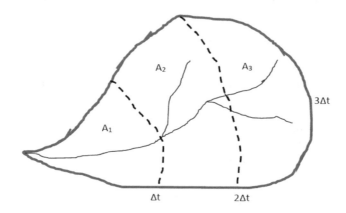

Solution:

Considering the calculation steps outlined in Table 5.3, the following results can be derived (where $\Delta t = 2$ min.):

Time	Discharge	Runoff contribution from			Discharge $(m^3/2\ min)$	Discharge (m^3/min)
		A_1	A_2	A_3		
Δt	Q_1	0.1*1	0	0	1.0	0.5
$2\Delta t$	Q_2	0.1*3	0.25*1	0	5.5	2.75
$3\Delta t$	Q_3	0.1*2	0.25*3	0.6*1	15.5	7.75
$4\Delta t$	Q_4	0	0.25*2	0.6*3	23.0	11.5
$5\Delta t$	Q_5	0	0	0.6*2	12.0	6.0

References

McDermott, G.E. and Pilgrim, D.H. (1982). Design Flood Estimations for Small Catchments in New South Wales. Australian Water Resources Council Technical Paper No. 73, ISBN: 0644 020741.

Pilgrim, D.H. (2001). Australian Rainfall and Runoff: A Guide to Flood Estimation, Vol. 1. Institution of Engineers Australia, Barton, ACT. ISBN: 0858257440.

Open Channel Hydraulics

6.1 Introduction

In an urban environment, runoff generated from individual small catchments is usually transported through stormwater pipes, although, in some cases, the runoff is directly transported to a nearby stream/creek. Even when it is transported through the stormwater pipes, it is eventually discharged into a nearby stream/creek. Often some of these creeks are converted to concrete-lined channels, mainly to avoid soil erosion from the natural surface. Figure 6.1 shows a photo of typical urban stormwater pipe outlet to a natural creek. For the hydraulic analysis of flow in open channels, relationships between discharge, water depth and velocity are often required.

Figure 6.1. A typical photo of stormwater pipe outlet to a natural creek

6.2 Principles and Equations

For the analysis of open channel flows, among the fluid mechanics principles, the following principles are frequently used:

- Conservation of mass
- Conservation of energy
- Hydrostatic pressure distribution
- Uniform velocity distribution

To explain these principles, consider a longitudinal section of a channel with flowing water, as shown in Figure 6.2. Often, two sections (1 and 2) are considered within a section of channel. Section 1 is referred as upstream section, while Section 2 is referred as downstream section.

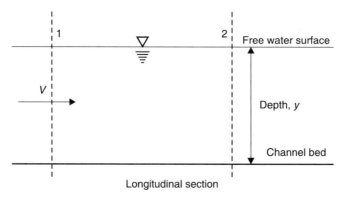

Figure 6.2. A typical longitudinal section of a channel with flowing water

'Conservation of mass' implies that transport of mass at 'section 1' will be same as transport of mass at 'section 2' at a given time period. In other words, amount of mass entering at 'section 1' (i.e. inflow) will be same as the amount of mass getting out from 'section 2' (i.e. outflow). As the amount of mass is equal to 'density*volume', for a particular fluid if the fluid density (between the considered sections) remains same (which is usually the case), then 'conservation of mass' also implies that, at a given time period, the volume of water/fluid passing through 'section 1' will be equal to the volume of water/fluid passing through 'section 2'. This eventually emerges as 'continuity equation', i.e. discharge at 'section 1' equals to the discharge at 'section 2', which can be expressed mathematically as, $Q_1 = Q_2$, where Q is the typical symbol for discharge.

'Conservation of energy' implies that energy at 'section 1' will be equal to the energy at 'section 2', provided that there is no addition/loss of energy in between. Energy at a section is calculated using 'energy equation' (i.e. Bernoulli's equation), explained later.

Such conservation analysis is typically done along the flow streamline, which is defined as the flow path of a fluid particle at a particular location/ depth. Also, while doing such analysis, it is assumed that the fluid/water in the channel has hydrostatic pressure, which implies: $P = \rho * g * y$, where, P is the pressure, y is the fluid depth, ρ is the fluid density and g is the acceleration due to gravity. Such assumption is valid for a channel which is horizontal (or close to horizontal, i.e. very mild slope).

For the analysis of an urban creek/channel, it is often assumed that the water velocity is uniform along the depth, as well as across the width of the channel. In reality, the velocity along the depth (as well as across the width) is not uniform, rather similar to the distribution as shown in Figure 6.3. However, such assumption of uniformity is acceptable for a shallow channel.

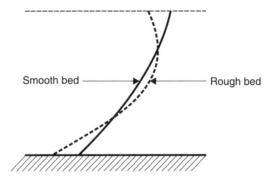

Figure 6.3. Typical velocity distribution along the depth in a channel

As mentioned earlier, a uniform velocity distribution is often assumed. This uniform velocity is called 'mean velocity', which can be calculated if the discharge and flow cross-sectional area of the channel are known. The following equation expresses a fundamental fluid mechanics relationship:

$Q = A*V$, where Q is the discharge, A is the cross-sectional area of the flow and V is the mean velocity of the flow. Through such a relationship, the earlier mentioned continuity equation (i.e. $Q_1 = Q_2$) can be written as:

$$A_1 V_1 = A_2 V_2 \tag{6.1}$$

This equation implies that for a flow within a channel, if the area of flow decreases, then the velocity has to increase in order to maintain the constant Q, and vice versa.

Similarly, conservation of energy implies that $E_1 = E_2$, where E is the specific energy at a section, which is defined as energy per unit weight of the waterbody at any section of the channel measured with respect to the channel bottom. Specific energy is expressed with the following equation:

$$E = \frac{P}{\rho g} + \frac{V^2}{2g} + Z \tag{6.2}$$

All other parameters are defined earlier in this section, except Z, which is the elevation or 'elevation head'. In this equation, the first term ($P/\rho g$) is termed as 'pressure head' and the second term ($V^2/2g$) is termed as 'velocity head'. As it is the energy per unit weight, the ultimate unit for 'specific energy' becomes a unit of 'length' (i.e. metre or centimetre). The pressure head is basically the water/fluid height within the channel, which exerts hydrostatic pressure. Therefore, the higher the water within the channel, the greater the pressure head will be. 'Velocity head' is the velocity of water/fluid, which contains dynamic energy. Thus, the higher the water velocity in the channel, the larger the velocity head. 'Elevation head' is the energy obtained/lost through a change in the elevation of water body/channel. As such, two channels might have water flowing with the same velocity, height and energy, however, one flowing with a higher elevation (i.e. bed level) will have a higher amount of energy. Also, these energy components are convertible to each other. For example, a container is holding water in it, as water is in a static condition, there is no velocity (i.e. velocity head is zero). However, the water in the container has a pressure head, which is being exerted to the container wall. Also, it has an elevation head, depending on the height/position of the container. Now, if a hole is made in the container, water will start to flow through the hole. In this case, some of the pressure head will convert into velocity head and the pressure head in the container will drop down. Similarly, as shown in Figure 6.4(a), the container in the right side will have a higher velocity through the hole compared to the velocity from the left container. This is because of the right side container having a higher pressure head, which is converting to velocity. Again, if the same container were raised to a higher level and water flowed through the same hole, the water would hit the ground wideth a higher velocity compared to the one which fell from the lower elevation. This higher velocity head is contributed from the additional elevation head attained through raising the container. Figure 6.4(b) shows such example, where water from the left side container will hit the ground with a higher velocity.

6.3 Effect of Streamline Position

Consider a flowing water through a horizontal rectangular channel, for which specific energy needs to be calculated. Here, assessment is being made while considering the flow streamline either at the channel bottom or at the water surface. Being a rectangular channel, one more parameter, discharge per unit width (W) can be derived, which is expressed with q (i.e. $q = Q/W$). Discharge can be expressed as $V*y*W$ and q can be expressed as

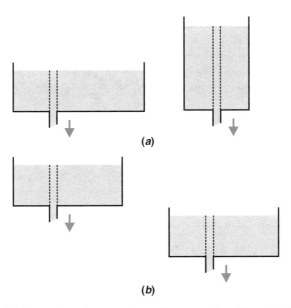

Figure 6.4. Examples of conversions of pressure head to velocity head

$V*y$. Considering streamline along the channel bed (Figure 6.5), pressure head ($P/\rho g$) for this scenario would be y ($=\rho g y/\rho g$) and velocity (V) could be expressed as q/y. As the specific energy is calculated with respect to the channel bottom, if the channel bottom is considered as the datum, then the elevation head is 'zero' (i.e. $Z=0$). Adding all these heads in to the energy equation, provides a final specific energy of:

$$E = \frac{P}{\rho g} + \frac{V^2}{2g} + Z = y + \frac{q^2}{2gy^2} + 0 = y + \frac{q^2}{2gy^2} \tag{6.3}$$

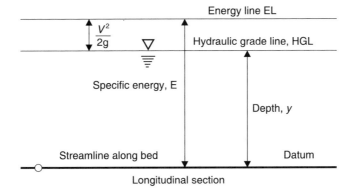

Figure 6.5. Specific energy considering streamline along the bed

Now, considering streamline along the water surface (Figure 6.6), pressure head ($P/\rho g$) for this scenario would be '0' (as there is no hydrostatic pressure on the surface) and velocity (V) can be expressed as q/y. From the channel bottom (i.e. datum) the elevation head is y (i.e. $Z=y$). Adding all these heads in to the energy equation, provides a final specific energy of:

$$E = \frac{P}{\rho g} + \frac{V^2}{2g} + Z = 0 + \frac{q^2}{2gy^2} + y = y + \frac{q^2}{2gy^2} \tag{6.4}$$

It is found that, in either consideration, the expression of specific energy remains the same.

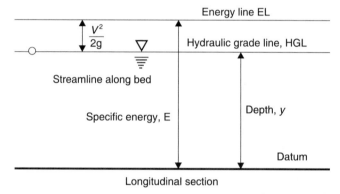

Figure 6.6. Specific energy considering streamline along the surface

6.4 Solutions of Energy Equation

If the mentioned energy equation is drawn for a set of E and y values for a particular q, the plot will be similar to the one shown in Figure 6.7, where the upper portion of the line will have a steep slope which eventually becomes 45° and the lower portion of the line will have a mild slope.

From the line, it is clear that for a particular E value, there are two possible depths (except for one particular point, the left extreme point having minimum E value on the plot), which are called alternate depths. The point having a higher depth is called subcritical depth, which will have a lower velocity, and the point having a lower depth is called supercritical depth, which will have a higher velocity. In between, there is a point, which is the crest of the curve having the minimum specific energy (E_{min}). Depth corresponding to the minimum specific energy is called 'critical depth (y_c)' and this minimum energy is referred to as 'critical energy' (Figure 6.8).

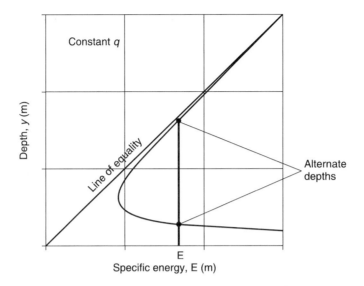

Figure 6.7. A typical plot of energy equation for a particular q

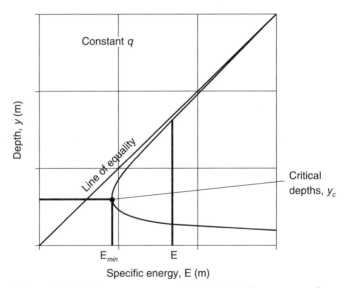

Figure 6.8. Minimum energy and critical depth on energy line

Through applying concepts of critical depth and alternate depths, several unknowns of open channel systems can be evaluated. For example, change of flow characteristics due to a change in bed level, a change in channel width, before and after a sluice gate, before and after a hydraulic jump and weir flow, which will be elaborated in the later sections.

It is to be noted that the energy equation can be expressed in three different formats:

$$E = y + \frac{V^2}{2g} = y + \frac{q^2}{2gy^2} = y + \frac{Q^2}{2gA^2} \tag{6.5}$$

Users select a particular format based on their needs and available known parameter(s). For example, the format with q is only valid for a rectangular channel. In an irregular channel, the q (i.e. discharge per unit width) is not constant (across the width) and, therefore, meaningless.

In regard to calculating a specific energy, if q and y values are known for a particular channel flow, then E can be calculated using the second format in Equation 6.5. However, if for channel flow E and q are known, while y is unknown, to calculate y there is no easy solution, rather the solution has to be reached through trials. Using the second format of the energy equation, the required steps for such a solution are outlined below:

- Step 1: Guess depth, y
- Step 2: Calculate E using the second format of the energy equation
- Step 3: Compare calculated E with the given value of E
- Step 4: Repeat steps 1 to 3 until the calculated E is the same (or very close) to the given value of E

In such solutions, there are likely to be two solutions for two alternate depths (one subcritical and the other supercritical), except for the case when the flow is critical (i.e. depth is critical depth). As such, users need to know their expected flow condition (subcritical/supercritical). Also, in some cases, there may be no solution, when available channel energy becomes less than the critical energy (i.e. $E < E_{min}$) for a particular flow condition. There is another trial method of finding alternate depth, which is claimed to be more efficient. This method rearranges the specific energy equation (with q) in the following format:

$$y = E - \frac{q^2}{2gy^2}$$

Then the following for such solutions are outlined below:

- Step 1: Guess depth, y
- Step 2: Substitute y in the right-hand side of the equation and calculate y in the left-hand side
- Step 3: Repeat step 2, until assumed y and calculated y are the same (or very close)

6.5 Critical Depth Calculations

Calculation of a critical depth is an important aspect for analysing open channel flows, as it differentiates the flow; any depth higher than the critical depth is a subcritical flow, whereas any depth smaller than critical depth is a supercritical flow. Critical depth occurs at a point where the energy is minimum (i.e. at the crest of the energy curve). At this point (i.e. the crest), the change of energy with respect to y would be zero. Differentiating the energy equation (with q) with respect to y, for a particular q, yields

$$\frac{dE}{dy} = \frac{d}{dy}\left(y + \frac{q^2}{2gy^2}\right) = 0, \text{ which yields,}$$

$$0 = 1 + \frac{q^2}{2g}\frac{d}{dy}(y^{-2})$$

$$\Rightarrow \qquad -1 = -\frac{2q^2}{2g}(y^{-3})$$

$$\Rightarrow \qquad \frac{q^2}{gy^3} = 1$$

$$\Rightarrow \qquad y = y_c = \sqrt[3]{\frac{q^2}{g}} \qquad\qquad (6.6)$$

Also, $\qquad\qquad q = \sqrt{gy_c^3} \qquad\qquad (6.7)$

Substituting, Equation 6.6 (i.e. critical flow) in to the energy equation (Equation 6.4),

$$E_{min} = E_c = y_c + \frac{q^2}{2gy_c^2} = y_c + \frac{gy_c^3}{2gy_c^2} = y_c + 0.5 * y_c = 1.5y_c \qquad (6.8)$$

These relationships are widely used for several open channel flow analyses. As the q is used (which is valid for rectangular channels only) in the above equations, these equations are valid only for rectangular channels.

For non-rectangular/irregular channels, a separate set of relationships are required. Considering an irregular channel (Figure 6.9), differentiating the energy equation having Q (Q is constant as per continuity) with respect to y,

$$\frac{dE}{dy} = \frac{d}{dy}\left(y + \frac{Q^2}{2gA^2}\right) = 0, \text{ which yields,}$$

$$0 = 1 + \frac{Q^2}{2g}\frac{d}{dy}(A^{-2})$$

$$\Rightarrow \qquad 0 = 1 + \frac{Q^2}{2g}\frac{d}{dA}(A^{-2})\frac{dA}{dy}$$

$$\Rightarrow \qquad 0 = 1 - \frac{2Q^2}{2g}(A^{-3})\frac{dA}{dy}$$

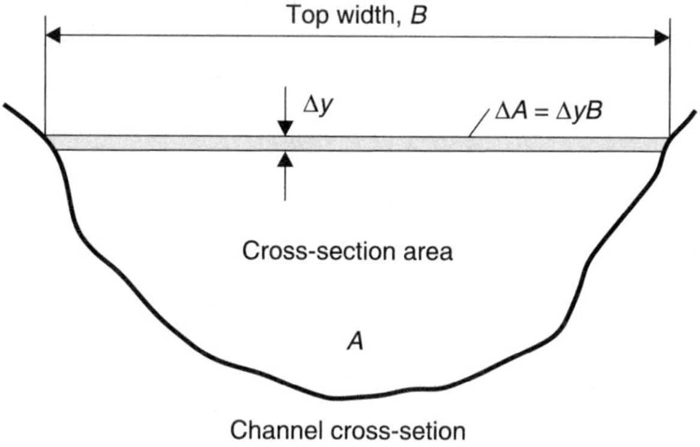

Top width, *B*

Δy   ΔA = ΔyB

Cross-section area

A

Channel cross-setion

Figure 6.9. Typical cross-section of an irregular channel

Considering a thin strip of the channel section at the top (Figure 6.9), dA/dy can be replaced with B (top width of the channel), which yields,

$$\Rightarrow \qquad 0 = 1 - \frac{Q^2}{g}(A^{-3})B$$

$$\Rightarrow \qquad \frac{Q^2 B}{gA^3} = 1 \qquad\qquad (6.9)$$

This critical flow relationship can be rearranged to derive the relationship with mean velocity (V), as follows:

$$\frac{Q^2 B}{gA^3} = 1$$

$$\frac{Q^2 B}{A^2 gA} = 1$$

$$\Rightarrow \qquad \frac{V^2}{g(A/B)} = 1, A/B \text{ is defined as hydraulic depth } (D), \text{ which yields,}$$

$$V^2 = gD \qquad\qquad (6.10)$$

6.6 Froude Number

Froude number is a dimensionless number, widely-used for the classification of the open channel flow. It basically expresses a ratio of inertia force to the gravity force for a certain open channel flow, where inertia force is assumed as the flow velocity (V) and gravity force is assumed as the hydraulic depth (D). To make it dimensionless, the ultimate Froude number (N_F) is defined as:

$$N_F = \frac{V}{\sqrt{gD}}$$

where D equals to A/B and for a rectangular channel, $D=A/B=y$.

Evaluation of Froude Number for Critical Flow

For a rectangular channel: For rectangular channel, $D=A/B=y=y_c$. Also, from Equation 6.7, $q = \sqrt{gy_c^3}$, substituting these into the equation for the Froude number,

$$N_F = \frac{V}{\sqrt{gD}} = \frac{q}{y_c\sqrt{gy_c}} = \frac{\sqrt{gy_c^3}}{y_c\sqrt{gy_c}} = \frac{\sqrt{gy_c^3}}{\sqrt{gy_c^3}} = 1$$

For an irregular channel: For irregular channel, $D=A/B$ and $V=Q/A$. Substituting these into the equation of the Froude number,

$$N_F = \frac{V}{\sqrt{gD}}, N_F^2 = \frac{V^2}{gD}$$

$$N_F^2 = \frac{V^2}{gD} = \frac{Q^2}{gA^2(A/B)} = \frac{Q^2B}{gA^3} = 1 \text{ (as per Equation 6.8)}$$

$$N_F = 1$$

So, for both the rectangular and irregular channels, for the critical flow, Froude number is 1. For subcritical flow, as the velocity is lower and the depth is higher, Froude number would be <1, and for supercritical flow, as the velocity is higher and the depth is lower, Froude number would be >1.

6.7 Applications of Energy Equation

In regard to the applications of energy equation, primarily the alternate depth can be calculated, i.e. for any hydraulic structure (or flow obstruction), i.e. depth of water before/after the structure can be calculated if one depth and q (or Q) are known. Weir, sluice gate, notch, change of the bed level and change of the channel width are a few such hydraulic structures. In addition to the evaluation of alternate depth before/ after hydraulic structures, energy equation can also be used to derive generalised equations for flows through the orifice, sluice gate, notch and weir. The following are a few examples of such derivations.

Orifice Flow

Consider an orifice flow under constant water head of depth y, as shown in Figure 6.10. Taking two points: 'point 1' on top of the water surface and 'point 2' just after the release of water through the orifice applying energy equations between these points yields.

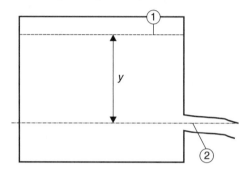

Figure 6.10. Flow through an orifice flow

At 'point 1', pressure head is 'zero' as it is under atmospheric pressure and the velocity head is also 'zero' as the water has a constant head (without any velocity). This point only has elevation head compared to the datum (the reference line passing through the orifice) and the elevation head, $Z=y$. So, total energy at 'point 1' is y. At 'point 2', the pressure head is 'zero' (i.e. atmospheric pressure), also the elevation head is 'zero' (as the point is just on the datum). The flow through the orifice has a velocity of V. So, the total energy at 'point 2' is $V^2/2g$. As per energy conservation, $E_1 = E_2$, which yields, $y = V^2/2g$, as such,

$$V = \sqrt{2gy} \tag{6.11}$$

Flow through Rectangular notch

Consider a flow through a rectangular notch having width b and a depth of flow H, as shown in Figure 6.11. Considering a strip of flow at a depth y having a thin strip thickness of dy. This thin strip can be approximated as an orifice with width b and thickness dy. So, the velocity (V) of flow through this strip would be as per Equation 6.10 and the area of flow is $dy*b$. As such, discharge (dQ) through this strip is:

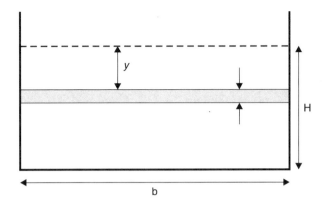

Figure 6.11. Flow through a rectangular notch

$$dQ = \sqrt{2gy} * b * dy$$

Now, integrating the above equation from 0 to H (i.e. for the whole depth of flow):

$$Q = \int_0^H dQ = b * \sqrt{2g} \int_0^H \sqrt{y}\,dy = b * \sqrt{2g} * \frac{H^{3/2}}{3/2}$$

$$Q = 2.96 * b * H^{3/2} \tag{6.12}$$

Flow through Triangular Notch

Consider a flow through a triangular 90° notch (i.e. side slope 1:1) having a depth of flow H, as shown in Figure 6.12. Considering a strip of flow at a depth y having a thin strip thickness of dy. This thin strip can be approximated as an orifice of thickness dy. As the side slopes are 1:1, and the depth below the strip is $(H - y)$, the width of the strip is $2(H - y)$. So, the velocity (V) of flow through this strip would be as per Equation 6.10 and the area of flow is $dy*2*(H - y)$. As such, discharge (dQ) through this strip is:

$$dQ = \sqrt{2gy} * 2(H - y) * dy = 2\sqrt{2g} * (H - y) * \sqrt{y} * dy$$

Now, integrating the above equation from 0 to H (i.e. for the whole depth of flow):

$$Q = \int_0^H dQ = 8.86 \int_0^H (H - y) * \sqrt{y}\,dy = 8.86 \int_0^H (H\sqrt{y}\,dy - y\sqrt{y}) * dy$$

$$Q = 8.86 * \left[\frac{H.H^{3/2}}{3/2} - \frac{H^{5/2}}{5/2}\right] = 2.363 * H^{5/2} \tag{6.13}$$

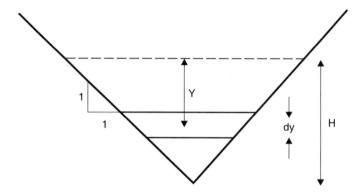

Figure 6.12. Flow through a triangular notch

Flow above a Weir

Usually, flow upstream of a weir is subcritical and flow above the weir is critical. Applying this concept, it is easy to derive an equation for the flow above a weir. Consider a weir flow having a depth of flow above the weir as y_c, as shown in Figure 6.13. Width of the weir is B.

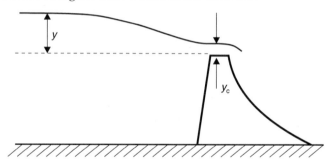

Figure 6.13. Flow above a weir

So, flow above the weir, $Q = A*V = B*y_c*V$

Being a critical flow, velocity on top of the weir can be written as, $V = \sqrt{gy_c}$

So, $Q = B*y_c*\sqrt{gy_c} = 3.13*B*y_c^{1.5}$

At critical flow, energy (E) can be written as equals to $1.5*y_c$, which yields,

$$Q = 3.13*B*\left(\frac{E}{1.5}\right)^{1.5} = 1.7*B*E^{1.5} \qquad (6.14)$$

Energy on top of the weir can be also evaluated by knowing the water depth (y) at far upstream, considering datum at the top of the weir (dashed line in the figure). At far upstream the water velocity can be approximated to 'zero' and in such a case the total energy of the flow would be y, which is equal to the energy on top of the weir, E_c (considering no energy loss between upstream and top of the weir). This sort of weir is typically called A 'broad crested weir' and previously discussed rectangular and triangular notches are typically called 'sharp crested weir'.

Flow through a Sluice Gate

Consider a flow (Q) through a sluice gate having an upstream depth of y_1 and downstream depth of y_2, as shown in Figure 6.14. Applying energy equation between 'section 1' (upstream) and 'section 2' (downstream) and considering no energy loss between the sections,

$$y_1 + \frac{V_1^2}{2g} = y_2 + \frac{V_2^2}{2g}$$

$$\Rightarrow \qquad y_1 + \frac{Q^2}{2gA_1^2} = y_2 + \frac{Q^2}{2gA_2^2}$$

(Note: A_1 and A_2 are the cross-sectional area of section 1 and section 2 respectively).

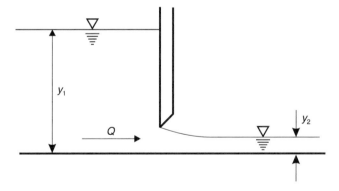

Figure 6.14. Flow through a sluice gate

$$\Rightarrow \qquad \frac{Q^2}{2g}\left[\frac{1}{A_1^2} - \frac{1}{A_2^2}\right] = y_2 - y_1$$

$$\Rightarrow \qquad Q^2\left[\frac{A_2^2 - A_1^2}{A_1^2 * A_2^2}\right] = 2g(y_2 - y_1)$$

$$\Rightarrow \qquad Q^2 = 2g(y_2 - y_1) * \left[\frac{A_1^2 * A_2^2}{A_2^2 - A_1^2} \right] \qquad (6.15)$$

If the width of the sluice gate is B, then the above equation becomes,

$$Q^2 = 2g(y_2 - y_1) * B^2 * \left[\frac{y_1^2 * y_2^2}{y_2^2 - y_1^2} \right] \qquad (6.16)$$

If y_1, y_2 and B are known for such flow, Q can be calculated using Equations 6.15 or 6.16.

In real life, if there is any energy loss within the sections, that should be deducted from the upstream energy. Also, if the bed level changes from upstream to downstream, the energy equations have to be adjusted based on the change of bed level, considering a certain datum (usually bottom of the channel).

Hydraulic Jump

Supercritical flow in nature is not really stable, as such, if there is a supercritical flow it often turns into a subcritical flow after forming a jump, called hydraulic jump (Figure 6.15). Conservation of energy cannot be applied between upstream and downstream of the jump, as there is significant energy loss during the formation of the jump.

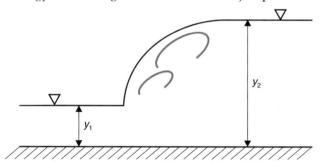

Figure 6.15. A typical sketch of hydraulic jump

For a hydraulic jump, the energy equations (i.e. conservation of energy) can be written as,

$$E_1 - \Delta E = E_2 \qquad (6.17)$$

$$\Rightarrow \qquad y_1 + \frac{V_1^2}{2g} - \Delta E = y_2 + \frac{V_2^2}{2g}$$

where ΔE can be calculated using the following relationship (Chow, 1973),

$$\Rightarrow \qquad \Delta E = \frac{(y_2 - y_1)^3}{4 * y_2 * y_1}$$

Also, if y_1 and Froude number at 'section 1' (F_1) are known then (Chow, 1973),

$$\frac{y_2}{y_1} = \frac{1}{2}\left(\sqrt{1 + 8 * F_1^2} - 1\right) \qquad (6.18)$$

Another term, jump height (h_j) is defined as, $y_2 - y_1$.

Change in Flow Characteristics due to Change in Bed Level

This is a very common channel characteristic change phenomena in urban waterway systems. Applying energy equation, the flow characteristics (i.e. depth, velocity) of the changed channel can be evaluated. Figure 6.16 shows a channel section, where the downstream channel bed is raised by a height of 'Δ', however, the total energy remains the same as the raising of bed level occurred smoothly, for which any loss of energy can be neglected. As the channel energy is measured with respect to the channel bottom, in such a scenario the downstream channel's energy will drop down by an amount of 'Δ', although the total energy remains the same as the upstream channel's total energy. If upstream channel energy is E_1 and downstream channel energy is E_2, then, $E_1 = E_2 + \Delta$.

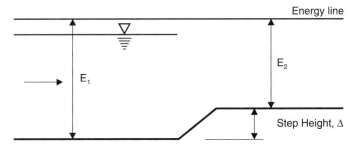

Figure 6.16. Typical channel section showing raise of bed level

In such a scenario (Figure 6.17), it is unknown whether the downstream water level will go down (dashed line) or go up (solid line). For such determination, further analysis is required.

Considering a rectangular channel having a bed level rise of Δ, the downstream profile can be determined through analysis of the energy line (Figure 6.18). For such a scenario, as E_2 would be smaller than E_1, the energy will shift towards the left on the curve depending on the magnitude of Δ. By moving along the energy line, if the upstream flow is subcritical (above the critical depth) then the downstream depth will go

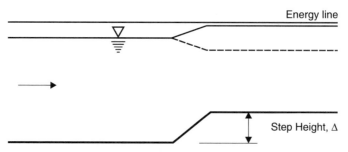

Figure 6.17. Typical channel section showing probable downstream water level

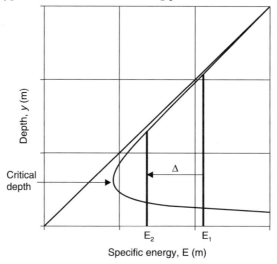

Figure 6.18. Determination of downstream profile through energy line

down and if the upstream flow is supercritical (below the critical depth) then the downstream depth will go up. Also, from the curve it is clear that the rate of dropping down (if occurs) would be steeper, while the rate of going up (if occurs) would be milder.

Actual determination of depth can be conducted using the energy equations in both the sections and applying the conservation of energy as, $E_1 = E_2 + \Delta$. In this case, as velocity (V) is changing from 'section 1' to 'section 2', whereas q remains the same for both the sections, the second form of energy equation has to be used:

$$y_1 + \frac{q^2}{2gy_1^2} = y_2 + \frac{q^2}{2gy_2^2} + \Delta \qquad (6.19)$$

In the above equation, q and Δ are usually known. Among y_1 and y_2, if one is known, then the other can be calculated through trial.

For another possible scenario of the lowering of a channel bed (Figure

6.19), similar analysis can be done. In this case, consider a rectangular channel, whose downstream bed is lowered by an amount Δ. Considering the downstream channel bed level as datum, the total energy of both the upstream and downstream channels can be equated as, $E_1 + \Delta = E_2$. Now, following the same steps as discussed in the previous case, if q and Δ are known, one depth can be calculated if the other depth is known. It is to be noted here that, once the depth is known, velocities can be calculated because q is known. For such a case, first it is necessary to determine whether the downstream water level will go up or down. As such, the solution requires trials. For quick convergence of the solution process, it is recommended that the solver knows whether the downstream water level will go up or down. This can be easily determined following the same procedure as discussed earlier in Figure 6.18. In this scenario, as E_2 would be higher than E_1, the energy will shift towards the right on the curve (Figure 6.20), depending on the magnitude of Δ. While moving along the energy line, if the upstream flow is subcritical (above the critical depth) then the downstream depth will go up and if the upstream flow is supercritical (below the critical depth) then the downstream depth will go down. Also, looking at the curve it is clear that the rate of dropping down (if occurs) would be milder, while the rate of going up (if occurs) would be steeper.

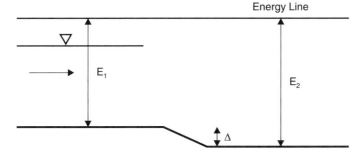

Figure 6.19. Typical channel section showing lowering of channel bed

Another fact that needs to be mentioned here is that the use of this particular energy equation is valid only for a rectangular channel. For irregular channels, similar determination can only be performed if the third format of the energy equation (i.e. using Q) is used, as in such case discharge (Q) remains the same for both the sections.

Change in Flow Characteristics due to Change in Channel Width

This is a common scenario in urban/natural channels and, due to a change in the channel width, flow characteristics (i.e. velocity, water level) also change. Energy equation can be applied in order to determine changed

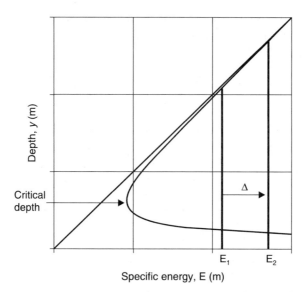

Figure 6.20. Determination of downstream profile through energy line

flow characteristics under such a scenario. For a rectangular channel, due to a change in the channel width, q changes, whereas Q remains the same. Also, if the channel bed level remains the same, then the energy will also remain the same, provided the change of width does not occur abruptly. For a particular channel section, if q increases, the energy curve shifts towards right, as shown in Figure 6.21, where q_2 is higher than q_1. As $q = \sqrt{g y_c^3}$, increase of q implies to increase of y_c and shifting the energy curve towards the right matches with the increase of y_c value.

Consider a reduction in channel width, as shown in Figure 6.22 (plan view). Considering total discharge Q, $q_1 = Q/W_1$ and $q_2 = Q/W_2$. As W_1 is higher than W_2, q_1 would be lower than q_2 and, as such, the energy curve for q_2 would shift towards the right.

To determine the possible profile of a downstream channel, first the energy of the current flow has to be determined. Then, drawing a vertical line at the particular energy (in x-axis of Figure 6.23) will reveal a possible change of the downstream profile from the intersections of the vertical line with the energy curves. Figure 6.23 shows the shifts of points (i.e. water depth); circular dots for q_1 and triangular dots for q_2. From the figure, it is clear that for the subcritical flow (i.e. depth above critical depth) the downstream water depth will go down and for the supercritical flow (i.e. depth below the critical depth) the downstream water depth will go up.

For the determination of downstream depth, the third format of the energy equation (as in this case Q is constant, not the q) should be used between section 1 and section 2. Applying conservation of energy:

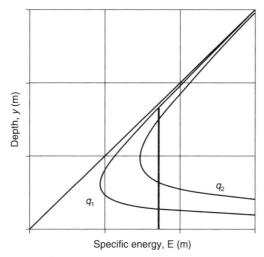

Figure 6.21. Shifting energy curve towards right due to increase in q

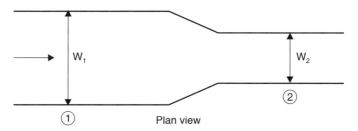

Figure 6.22. Reduction in channel width

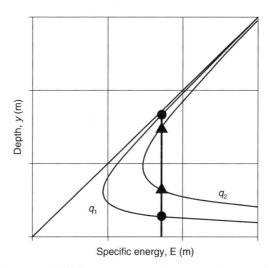

Figure 6.23. Shifting of water depths due to increase in q

$$y_1 + \frac{Q^2}{2gA_1^2} = y_2 + \frac{Q^2}{2gA_2^2} , \qquad (6.20)$$

which implies:

$$y_1 + \frac{Q^2}{2gW_1^2 y_1^2} = y_2 + \frac{Q^2}{2gW_2^2 y_2^2} \qquad (6.21)$$

Using the above equation, for a channel flow, if Q, W_1 and W_2 are known, then among the depths (y_1 and y_2), if one is known, then the other can be calculated through trials. The same equation can be used for an irregular channel, however in such a case as W_1 and W_2 has no meaning the previous equation (6.19) should be used.

If the reduction of width of a downstream channel is maintained, then at one stage the energy curve will shift too far away from the vertical energy line, as shown in Figure 6.24. In such a case, as the current energy line (vertical line) is not intersecting the shifted energy curve, mathematically no solution can be obtained for y_2, which physically means such reduction of a downstream channel width will cause a change of upstream flow characteristics. To keep the upstream flow condition unchanged, the maximum shifting of the energy curve is up to the vertical energy line (i.e. when the critical point of the energy curve just touches the vertical line, as shown in Figure 6.25). In such a case, the specific energy is the critical energy and the corresponding depth is the critical depth. For a particular channel having particular energy, critical depth can be calculated using $E = 1.5*y_c$. Then, from known y_c, q can be calculated using $q = \sqrt{gy_c^3}$ and this q is the maximum q in that particular channel which can be adopted while keeping the upstream flow conditions unchanged.

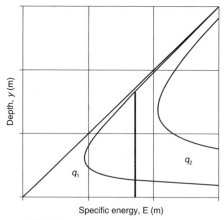

Figure 6.24. Shifting of the energy curve for very high q

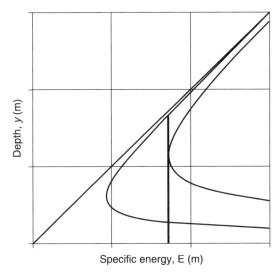

Figure 6.25. Energy curve just touching the current energy

6.8 Gravity Wave and its Applications

In hydraulics, gravity waves are waves generated in a fluid medium when the force of gravity tries to resist its flow or restore it back to equilibrium. A gravity wave results when fluid is displaced from a position of equilibrium/static. As minimum energy is spent when the flow moves as 'critical flow', such gravity flow tries to maintain a critical flow, although gravity force and subsequent friction tries to restrain its movement. As it moves with a critical velocity, considering a rectangular (or several slots of rectangular) channel(s), the velocity of such flow can be approximated as, $V = \sqrt{gy}$, where y is the depth of water (Lighthill, 2001).

Consider a gravity flow created by throwing a rock into the middle of a still pond, propagation of gravity waves is shown in Figure 6.26, where propagation is shown as circular lines and a solid circular dot at the centre is the location where the rock was thrown. As the velocity of these waves is $\sqrt{gy}$, if one can measure the velocity of this propagation, the depth of the water can be calculated from the above relationship.

Now, if the rock is thrown in a channel having a low velocity $V < \sqrt{gy}$), the propagation will be similar to the one shown in Figure 6.27 (consider that the channel has a velocity, V, towards the right). In this scenario, the waves towards the right will be moving faster than the waves towards the left. The resultant velocity towards the right is $V + \sqrt{gy}$ and towards the left is $\sqrt{gy} - V$. If one can measure these velocities, the depth y and velocity V can be calculated.

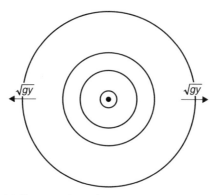

Figure 6.26. Propagation of gravity waves in a still pond

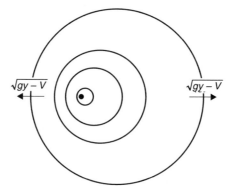

Figure 6.27. Propagation of gravity waves in a channel with low velocity

In another scenario, if the rock is thrown into a channel with a high velocity $V > \sqrt{gy}$), the propagation will be similar to the one shown in Figure 6.28 (consider that the channel has a velocity, V, towards the right). In this scenario, no wave will move towards the left, as the velocity/ movement of the gravity wave ($\sqrt{gy}$) towards the left will be surpassed by the high channel velocity opposite to the gravity wave towards left. The resultant velocity at the right boundary is $V + \sqrt{gy}$ and at the left boundary is $V - \sqrt{gy}$ (also moving towards the right).

Now, if the rock is thrown into a channel having a critical velocity (i.e. $V = \sqrt{gy}$), the propagation of waves will look like as shown in Figure 6.29. Since the channel (rectangular) has a critical velocity, the resultant wave velocity towards the right would be $2\sqrt{gy}$ and the resultant velocity towards the left would be 0 $(-\sqrt{gy} - \sqrt{gy})$. As such, by observing the pattern of the wave propagation following a rock fall in the flowing channel (rectangular or close to rectangular), one can determine the flow regime (i.e. critical/subcritical/supercritical).

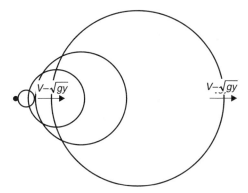

Figure 6.28. Propagation of gravity waves in a channel with high velocity

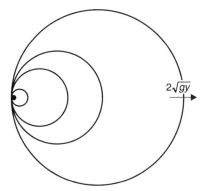

Figure 6.29. Propagation of gravity waves in a channel with high velocity

Worked Example 1

Water is flowing above a weir having a top level of 31 m AHD (Australian Height Datum). At the far upstream approaching water is flowing with a velocity of 0.005 m/s with water surface level of 31.4 m AHD. Determine the depth and velocity of the water above the weir. If the width of the weir is 12 m, then determine the discharge above the weir.

Solution:

Here, $V=0.005$ m/s and $y=31.4 - 31.0=0.4$ m. Using the first format of the energy equation (6.4), total energy of the approaching flow = $y + \dfrac{V^2}{2g} = 0.4$ + $0.005^2/(2*9.81) = 0.4$ m

So, energy on top of weir also 0.40 m and critical flow occurs on top of weir. So, depth of water on top of weir, y_c = Energy/1.5 = 0.4/1.5 = 0.267 m and velocity = $\sqrt{gy_c} = 1.62$ m/s.

Discharge above the weir, Q = velocity*flow area = 1.62*(0.267*12) = 5.185 m³/s.

Worked Example 2

Water is flowing through an irregular channel and, at one section, a critical flow is formed. At the critical section, top-width of the flow is 4.50 m, with a cross-sectional area of 3.0 m². Farther downstream, the channel shape has been changed to a rectangle having a width of 3.0 m. If another critical flow condition is formed at this rectangular section, then what is the velocity of water at this section?

Solution:

For an irregular channel, $\dfrac{Q^2 B}{g A^3} = N_F^2$

For critical flow, $N_F = 1$. So, $\dfrac{Q^2 B}{g A^3} = 1$

Given that, $B = 4.5$ m, $A = 3.0$ m²

From the above relationship, $Q^2 = \dfrac{g A^3}{B} = \dfrac{9.81 * 3^3}{4.5} = 58.86$

So, $Q = 7.672$ m³/s

At downstream rectangular section, width $(W) = 3.0$ m

$$q = Q/W = 7.672/3.0 = 2.557$$

For critical flow in rectangular section, $y_c = \sqrt[3]{\dfrac{q^2}{g}} = \sqrt[3]{\dfrac{2.557^2}{9.81}} = 0.874$ m

Velocity $(V) = q/y_c = 2.557/0.874 = 2.93$ m/s

Worked Example 3

Water is flowing through an irregular channel and, at one section, critical flow is formed. At the critical section, top-width of the flow is 5.0 m. At an upstream section, the channel shape was a rectangle having a width of 4.0 m. Another critical flow condition was formed at this rectangular section, having a critical depth of 0.80 m. Determine the cross-sectional area and average velocity of water at the irregular section.

Solution:

At the rectangular section for critical depth, $y_c = \sqrt[3]{\dfrac{q^2}{g}} = 0.80$ m

So, $q = \sqrt{g y_c^3} = \sqrt{9.81 * 0.8^3} = 2.24$

Discharge, $Q = q*W = 2.24*4.0 = 8.965$ m³/s

For a critical flow in an irregular section, $\frac{Q^2 B}{g A^3} = 1$, which yields,

$$A^3 = \frac{Q^2 B}{g} = \frac{8.965^2 * 5.0}{9.81} = 40.96$$

So, $A = 3.45 \text{ m}^2$

Average velocity, $V = Q/A = 8.965/3.45 = 2.60 \text{ m/s}$

Worked Example 4

Water is flowing through a rectangular channel and at a downstream section (channel width 5.0 m) a critical flow is formed with a depth of 0.75 m. At an upstream section (rectangular), the channel width was 4.0 m and another critical flow condition was formed. Determine the depth and velocity of water flow at this upstream section.

Solution:

Given: y_c at downstream section = 0.75 m, width at downstream section (W_d) = 5.0 m and width at upstream section (W_u) = 4.0 m.

Considering critical flow at downstream section,

$$q_d = \sqrt{g y_c^3} = \sqrt{9.81*0.75^3} = 2.035 \text{ m}^2/\text{s}.$$

So, discharge $(Q) = q_d * W_d = 2.035*5.0 = 10.17 \text{ m}^3/\text{s}.$

Now, q in upstream section, $q_u = Q/W_u = 10.17/4.0 = 2.543$ and

Upstream $\qquad y_c = \sqrt[3]{\frac{q_u^2}{g}} = 0.87 \text{ m}$

So, velocity at upstream = $q_u/(\text{upstream } y_c) = 2.543/0.87 = 2.92 \text{ m/s}.$

Worked Example 5

Water flows in a rectangular channel which is 2.0 m wide. A hydraulic jump occurred along the channel downstream and has the profile shown in the figure below. The depths upstream and downstream of the jump are measured at 0.25 m and 1.8 m, respectively. Calculate the flow rate in the channel, assuming that the loss of energy through occurrence of the hydraulic jump is 0.5 m.

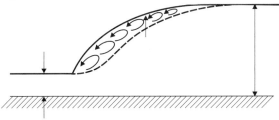

Solution:

Given, $W = 2.0$ m, $y_1 = 0.25$ m, $y_2 = 1.8$ m and $\Delta E = 0.5$ m

Applying Equation 6.16, $E_1 - \Delta E = E_2$, which yields,

$$y_1 + \frac{q^2}{2gy_1^2} - \Delta E = y_2 + \frac{q^2}{2gy_2^2}$$

$$\Rightarrow \qquad \frac{q^2}{2g}\left(\frac{1}{y_1^2} - \frac{1}{y_2^2}\right) = y_2 + \Delta E - y_1 = 1.80 + 0.5 - 0.25 = 2.05$$

$$\Rightarrow \qquad \frac{q^2}{2g} = 2.05 * \left(\frac{y_1^2 * y_2^2}{y_2^2 - y_1^2}\right) = 0.131$$

$$q^2 = 2*9.81*0.131 = 2.563$$

$$q = 1.60 \text{ m}^2/\text{s}$$

$$Q = q*W = 1.60*2 = 3.20 \text{ m}^3/\text{s}$$

Worked Example 6

Water flows in a rectangular channel which is 2.2 m wide. A hydraulic jump occurred along the channel downstream and has the profile shown in the figure below. The depths upstream and downstream of the jump are measured at 0.20 m and 1.70 m, respectively. Calculate the flow rate in the channel, assuming that the loss of energy through occurrence of the hydraulic jump is 0.5 m.

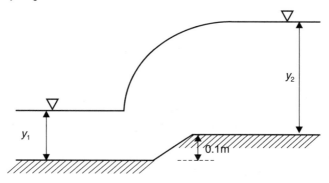

Solution:

Given, $W = 2.2$, $y_1 = 0.20$ m, $y_2 = 1.7$ m, $\Delta = 0.1$ m and $\Delta E = 0.5$ m

Applying Equation 6.16 (considering bed level of upstream section as datum), $E_1 - \Delta E = E_2 + \Delta$, which yields,

$$y_1 + \frac{q^2}{2gy_1^2} - \Delta E = y_2 + \frac{q^2}{2gy_2^2} + \Delta$$

$$\Rightarrow \qquad \frac{q^2}{2g}\left(\frac{1}{y_1^2} - \frac{1}{y_2^2}\right) = y_2 + \Delta E - y_1 + \Delta = 1.7 + 0.5 - 0.2 + 0.1 = 2.1$$

$$\Rightarrow \qquad \frac{q^2}{2g} * 24.65 = 2.1$$

$$\Rightarrow \qquad q^2 = 2.1 * 2.0 * 9.81 / 24.65 = 1.67$$

So, $q = 1.292$ and $Q = q*W = 1.292*2.2 = 2.842 \ \mathrm{m^3/s}$

Worked Example 7

Water flows in a rectangular channel which is 2.0 m wide. The depth of water in the channel is controlled by a sluice gate, as shown in the figure below. The depths upstream and downstream of the gate are measured at 2.0 m and 0.20 m, respectively. Calculate the flow rate in the channel, assuming that the specific energy is conserved as the flow goes through the opening.

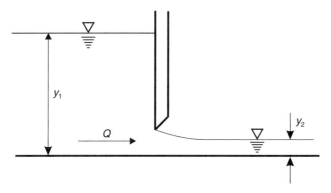

Solution:

Given, $W = 2.0 \ \mathrm{m}$, $y_1 = 2.0 \ \mathrm{m}$ and $y_2 = 0.20 \ \mathrm{m}$

Applying the energy equation in both the sections and conservation of energy,

$$y_1 + \frac{q^2}{2gy_1^2} = y_2 + \frac{q^2}{2gy_2^2}$$

$$\Rightarrow \qquad \frac{q^2}{2g}\left(\frac{1}{y_1^2} - \frac{1}{y_2^2}\right) = y_2 - y_1$$

$$\Rightarrow \qquad \frac{q^2}{2g} = (y_2 - y_1) * \left(\frac{y_1^2 * y_2^2}{y_2^2 - y_1^2}\right) = (y_2 - y_1) * \left(\frac{y_1^2 * y_2^2}{(y_2 - y_1) * (y_2 + y_1)}\right)$$

$$\Rightarrow \qquad q^2 = 2g * \left(\frac{y_1^2 * y_2^2}{y_2 + y_1} \right) = 1.427$$

So, $q = 1.195$ and $Q = q*W = 1.195*2.0 = 2.39$ m³/s

Worked Example 8

Water flows in a rectangular channel which is 2.2 m wide. Water passed beneath a sluice gate and then a hydraulic jump occurred at the channel downstream, the profile of which is shown in the figure below. The depth upstream of the gate, y_1 is 1.75 m and downstream of the jump, y_3 is 1.50 m. Calculate the flow rate in the channel, assuming that the loss of energy through occurrence of the hydraulic jump is 0.15 m. There is no energy loss between the upstream and downstream of the gate.

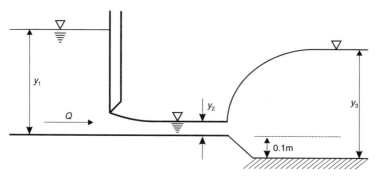

Solution:

Given, $W = 2.2$, $y_1 = 1.75$ m, $y_3 = 1.5$ m, $\Delta = 0.1$ m and $\Delta E = 0.15$ m

Applying Equation 6.16 (considering bed level of downstream section as datum), $E_1 + \Delta - \Delta E = E_3$, which yields

$$y_1 + \frac{q^2}{2gy_1^2} + \Delta - \Delta E = y_3 + \frac{q^2}{2gy_3^2}$$

$$\Rightarrow \qquad \frac{q^2}{2g} \left(\frac{1}{y_1^2} - \frac{1}{y_3^2} \right) = y_3 - y_1 - \Delta + \Delta E = 1.5 - 1.75 - 0.1 + 0.15$$

$$\Rightarrow \qquad \frac{q^2}{2g} \left(\frac{1}{1.75^2} - \frac{1}{1.5^2} \right) = -0.20$$

$$\Rightarrow \qquad \frac{q^2}{2g} \left(\frac{1}{1.75^2} - \frac{1}{1.5^2} \right) = -\frac{0.20}{-0.12} = 1.69$$

$$q^2 = 2*9.81*1.69 = 33.16$$

$$q = 5.76$$

So, $Q = q*W = 5.76*2.2 = 12.67 \text{ m}^3/\text{s}.$

Worked Example 9

Water is flowing over a broad-crested weir, forming a critical depth (0.132 m) above the weir. The width (B) of the weir is 15.0 m (weir equation can be applied in order to calculate discharge in the downstream channel). Then, it flows far downstream to a rectangular channel that has a width of 2.25 m and a depth of 1.0 m. At farther downstream sections, the bed of the channel is first raised by 100 mm and then raised by another 100 mm (as shown in the figure below). It can be considered that the 'conservation of energy' principle applies along the downstream channel sections. Determine the depth of flow in the most downstream section. You can assume that there are no energy losses within the sections.

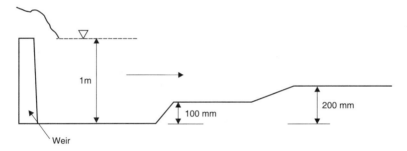

Solution:

Given, $y_c = 0.132$ m, $B = 15.0$ m, $\Delta = 0.1+0.1 = 0.20$ m and $W = 2.25$ m

Energy on top of weir (i.e. critical energy), $H = 1.5*y_c = 1.5*0.132 = 0.198$ m

Applying weir equation, $Q = 1.7*B*H^{1.5}$

Discharge, $Q = 1.7*15*0.198^{1.5} = 2.25 \text{ m}^3/\text{s}$

Consider, 'section 1' is the section immediately downstream of the weir and 'section 2' is the far downstream section. (Note: For this analysis we do not need to consider the middle section of the channel.)

Velocity at section 1 of the channel, $V_1 = Q/A = 2.25/(2.25*1) = 1$ m/s.

$$q_1 = V_1*y_1 = 1*1 = 1 \text{ m}^2/\text{s}$$

$$q_2 = q_1$$

Energy at section 1, $E_1 = y_1 + \dfrac{V_1^2}{2g} = 1 + \dfrac{1^2}{2g} = 1.051\text{m}$

$$E_2 = E_1 - \Delta = 1.051 - 0.20 = 0.851 \text{ m}$$

Froude number at section 1, $N_F = \dfrac{V_1}{\sqrt{gy_1}} - \dfrac{1}{\sqrt{9.81}} = 0.32$

As the Froude number is less than '1', the flow is subcritical. So, as per the discussed theory, the downstream water level will drop down. Expressing energy equation for the 'section 2':

$$E_2 = 0.851 = y_2 + \frac{q_2^2}{2g*y_2^2} = y_2 + \frac{0.051}{y_2^2}$$

By trial, $y_2 = 0.76$ m

Worked Example 10

Water is flowing over a broad-crested weir, forming a critical depth (0.132 m) above the weir. The width (B) of the weir is 15.0 m (weir equation can be applied in order to calculate discharge in the downstream channel). Then, it flows far downstream to a rectangular channel that has a width of 2.25 m and forming a depth of 0.8 m. At farther downstream sections, the bed of the channel is first raised by 100 mm and then dropped by 200 mm (as shown in the figure below). It can be considered that the 'conservation of energy' principle applies along the downstream channel sections. Determine the depth of flow in the most downstream section. You can assume that there are no energy losses within the sections.

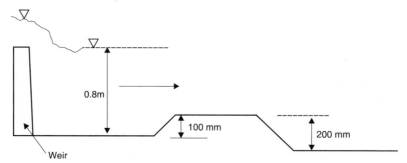

Solution:

Given, $y_c = 0.132$ m, $B = 15.0$ m and $W = 2.25$ m

Energy on top of weir (i.e. critical energy), $H = 1.5*y_c = 1.5*0.132 = 0.198$ m

Applying weir equation, $Q = 1.7*B*H^{1.5}$

Discharge, $Q = 1.7*15*0.198^{1.5} = 2.25$ m³/s

Consider, 'section 1' is the section immediately downstream of the weir and 'section 2' is the far downstream section. (Note: For this analysis we do not need to consider the middle section of the channel.)

Considering the downstream channel bed level as datum, $\Delta = 0.2 - 0.1 = 0.1$ m.

Velocity at section 1 of the channel, $V_1 = Q/A = 2.25/(2.25*0.8) = 1.25$ m/s.

$$q_1 = V_1 * y_1 = 1*0.8 = 1 \text{ m}^2/\text{s}$$

$$q_2 = q_1$$

Energy at section 1, $E_1 = y_1 + \dfrac{V_1^2}{2g} = 0.8 + \dfrac{1.25^2}{2g} = 0.88\text{m}$

$$E_2 = E_1 + \Delta = 0.88 + 0.10 = 0.98 \text{ m}$$

Froude number at section 1, $N_F = \dfrac{V_1}{\sqrt{gy_1}} = \dfrac{1.25}{\sqrt{9.81*0.8}} = 0.45$

As the Froude number is less than '1', the flow is subcritical. So, as per the discussed theory, the downstream water level will go up. Expressing energy equation for the 'section 2':

$$E_2 = 0.98 = y_2 + \frac{q_2^2}{2g * y_2^2} = y_2 + \frac{0.051}{y_2^2}$$

By trial, $y_2 = 0.92$ m

Worked Example 11

Water flows at a depth of 1.60 m in a 3.50 m wide rectangular channel, at a velocity of 0.625 m/s. At a section downstream, the channel bed is raised by 325 mm. Calculate the depth of flow at the section downstream and the minimum specific energy for this flow rate in this channel. Also, determine how much the channel bed (at downstream) can be raised without affecting the flow in the upstream.

Solution:

Given, $y_1 = 1.60$ m, $W = 3.5$ m, $V_1 = 0.625$ m/s and $\Delta = 0.325$

Energy at section 1, $E_1 = y_1 + \dfrac{V_1^2}{2g} = 1.6 + \dfrac{0.625^2}{2g} = 1.62\text{m}$

$$E_2 = E_1 - \Delta = 1.62 - 0.325 = 1.295 \text{ m}$$

$$q = y_1 * V_1 = 1.0$$

Froude number at section 1, $N_F = \dfrac{V_1}{\sqrt{gy_1}} = \dfrac{0.625}{\sqrt{9.81*1.6}} = 0.158$

As Froude number less than 1, the flow is subcritical and the water level will go down.

Now,
$$E_2 = 1.295 = y_2 + \frac{q^2}{2g * y_2^2} = y_2 + \frac{0.051}{y_2^2}$$

By trial, $y_2 = 1.26$ m

Critical depth for the flow, $y_c = \sqrt[3]{\frac{q^2}{g}} = 0.87$ m $= 0.467$ m and the minimum specific energy, $E_c = 1.5*y_c = 0.70$ m.

Downstream channel bed can be raised (without affecting the upstream flow profile) up to the point when $E_2 = E_c$. So maximum allowable $\Delta = E_1 - E_c = 1.62 - 0.70 = 0.92$ m

Therefore, the downstream bed level can be raised to 0.92 m without affecting the upstream flow profile.

Worked Example 12

Water flows at a depth of 2.50 m in a 3.25 m wide rectangular channel, at a velocity of 0.354 m/s. At a section downstream, the channel width is reduced to 2.80 m. Calculate the depth of flow at the section downstream. Also, calculate the maximum possible discharge per unit width for this specific energy in this channel without affecting the upstream flow profile.

Solution:

Given, $y_1 = 2.5$ m, $W_1 = 3.25$ m, $V_1 = 0.354$ m/s and $W_2 = 2.80$ m

Energy at section 1, $E_1 = y_1 + \frac{V_1^2}{2g} = 2.5 + 0.354^2/(2*9.81) = 2.506$ m

Froude number at section 1, $N_F = \frac{V_1}{\sqrt{gy_1}} = \frac{0.354}{\sqrt{9.81*2.5}} = 0.07$

As the Froude number is less than '1', the flow is subcritical and, as such, flow in the downstream section will drop down.

In the current scenario, $E_1 = E_2$ and Q remains constant.

$$Q = y_1*W_1*V_1 = 2.876 \text{ m}^3/\text{s}$$

Applying the third format of the energy equation considering Q,

$$y_1 + \frac{Q^2}{2gA_1^2} = y_2 + \frac{Q^2}{2gA_2^2}$$

$$\Rightarrow \qquad y_1 + \frac{Q^2}{2gy_1^2W_1^2} = y_2 + \frac{Q^2}{2gy_2^2W_2^2}$$

$$\Rightarrow \qquad 2.5 + \frac{2.876^2}{2*9.81*2.5^2*3.25^2} = y_2 + \frac{2.876^2}{2*9.81*y_2^2*2.8^2}$$

$$\Rightarrow \qquad 2.506 = y_2 + \frac{0.0538}{y_2^2}$$

By trial, $y_2 = 2.495$ m.

If we keep on increasing q, the energy curve will keep on shifting towards to right and, at one stage, the energy curve will just touch the current energy line (E_1) and that's the point when the current energy becomes critical energy (E_c). Beyond this point, there will be no solution and it will affect the upstream flow profile. So, considering the current energy as the critical energy,

$$E_c = 2.506 \text{ and } y_c = 2/3*E_c = 1.671 \text{ m}$$

Hence, maximum allowable discharge per unit width,

$$q_{max} = \sqrt{gy_c^3} = \sqrt{9.81*1.671^3} = 6.766 \text{ m}^2/\text{s}.$$

Worked Example 13

Water flows at a depth of 0.5 m in a 2.0 m wide rectangular channel, at a velocity of 2.5 m/s. At a section downstream, it became necessary to shorten the channel width. Is it possible to shorten the width to 1.0 m, without changing the upstream profile depth? If yes, calculate the depth of flow at the downstream section. If no, why?

Solution:

Given, $y_1 = 0.5$ m, $W_1 = 2.0$ m and $V_1 = 2.5$ m/s.

Discharge, $Q = 2.5*0.5*2 = 2.5 \text{ m}^3/\text{s}$

And, energy, $E_1 = y_1 + \dfrac{V_1^2}{2g} = 0.5 + 2.5^2/(2*9.81) = 0.82$ m

For a minimum possible width (without affecting upstream flow profile), at least a critical section must be created. So considering, $E_c = E_1 = 0.82$ m

So, $y_c = E_c/1.5 = 0.55$ m and discharge per unit width for this E_c,

$$q_{max} = \sqrt{gy_c^3} = \sqrt{9.81*0.55^3} = 1.278 \text{ m}^2/\text{s}$$

So, minimum possible width = $Q/q_{max} = 2.5/1.278 = 1.96$ m

Therefore, it is not possible to shorten the channel width to 1.0 m, as the minimum possible width for the downstream channel is 1.96 m.

144 Urban Water Resources

Worked Example 14

Water flows in a 2.25 m wide rectangular channel at a depth of 1.60 m and has a velocity of 0.850 m/s. At a section downstream, the bed of the channel needs to be raised by 1000 mm. Will it affect the upstream water profile? If yes, what modifications in the channel can be proposed to keep the upstream water profile unchanged? You can assume that there are no energy losses between the two sections.

Solution:

Given, $y_1 = 1.6$ m, $W_1 = 2.25$ m and $V_1 = 0.85$ m/s.

Discharge, $Q = y_1*W_1*V_1 = 3.06$ m³/s

Discharge per unit width, $q = Q/W_1 = 1.36$ m²/s

Energy, $E_1 = y_1 + \dfrac{V_1^2}{2g} = 1.6 + 0.85^2/(2*9.81) = 1.64$ m

With the proposed raising,

$$E_2 = E_1 - \Delta = 1.64 - 1.0 = 0.64 \text{ m}$$

With the current q, critical depth, $y_c = \sqrt[3]{\dfrac{q^2}{g}} = 0.574$ m and critical energy, $E_c = 1.5*y_c = 0.86$ m.

As E_2 is less than E_c, the proposed raising of be level will affect the upstream water profile. To keep the upstream profile unchanged, we need to increase the width of the channel (i.e. to decrease the q) and we have to at least create a critical section at 'section 2'. So, new $E_c = E_2 = 0.64$

New, $y_c = E_c/1.5 = 0.43$ m

$$q = \sqrt{gy_c^3} = \sqrt{9.81*0.43^3} = 0.872 \text{ m}^2/\text{s}$$

So, proposed width, $W_2 = Q/q = 3.06/0.872 = 3.51$ m

So, to keep upstream profile unchanged the downstream channel width has to be enlarged to a minimum width of 3.51 m.

Worked Example 15

Water is flowing south in a channel. A stone is thrown into in the middle of the channel. Waves created by the stone moved at a rate 10 cm/s towards the north and 200 cm/s towards the south. Determine the depth of the channel at the location where the stone was thrown. Also, determine whether the flow in the channel is subcritical or supercritical.

Solution:

From the given data,

$$V + \sqrt{gy} = 2.0$$

$$V - \sqrt{gy} = -0.10$$

From the above two equations, $2V = 1.90$
So, $V = 0.95$
Now, from the relation, $V + \sqrt{gy} = 2.0$, $\sqrt{gy} = 1.05$
So, $gy = 1.1025$, $y = 0.1124$ m

Froude number, $N_F = \dfrac{V}{\sqrt{gy}} = \dfrac{0.95}{\sqrt{9.81*0.1124}} = 0.905$

As the Froude number is less than '1', the flow is subcritical.

References

Chow, V.T. (1973). Open Channel Hydraulics. McGraw Hill Company Japan, ISBN 007 085906X.

Lighthill, M.J. (2001). Waves in Fluids. Cambridge University Press, p. 205, ISBN 9780521010450.

Uniform Flow in Open Channel

7.1 Flow Classifications

In addition to the flow classifications discussed in the previous chapter (i.e. critical/subcritical/supercritical flows), some more classifications are frequently used for open channel flows. Among those classifications, the two most frequently used ones are mentioned below:

Steady/Unsteady Flow

If the basic flow properties (i.e. depth of water and velocity/discharge) in a channel section remain the same with respect to time, then it is called steady flow, and if any of those properties changes with time then it is called an unsteady flow. Unsteady flow equations are complicated and are generally used for very complicated and long river networks. In general, for urban channel/creek networks, if a small segment of a channel/creek is analysed, then the steady flow equations are good enough in most cases. In reality, considering a longer time period, none of the flows are steady. However, to take advantage of simpler analysis using steady flow equations, smaller time periods are considered, during which a particular flow may be considered as steady. In this book, only the steady flow equations will be elaborated. Mathematically, steady flow properties can be defined as:

$$\frac{\partial y}{\partial t} = 0 \text{ and } \frac{\partial V}{\partial t} = 0$$

where y is the water depth and V is the flow velocity.

Uniform/Non-uniform Flow

If the basic flow properties (i.e. depth of water and velocity/discharge) within a channel section remain the same with respect to its horizontal distance (often referred as x), then it is called a uniform flow and if any of those properties changes with x, then it is called a non-uniform flow. Mathematically, steady flow properties can be defined as:

$$\frac{\partial y}{\partial x} = 0 \text{ and } \frac{\partial V}{\partial x} = 0$$

where x is the direction of flow (i.e. horizontal for very mild channel).

In reality, for a whole channel/creek network the flows are not likely to be uniform, as within the flow path of channels/creeks there are usually many hydraulic structures/obstructions. These structures/obstructions cause a flow to change from uniform to non-uniform. However, within certain sections of a channel/creek, where channel properties (i.e. cross-section, slope) remain the same, a uniform flow can be assumed and, through such assumption, several flow properties/unknowns can be analysed. These will be discussed in the following sections. In this book, only the uniform flow equation will be elaborated.

Practical Examples on Formation of Uniform Flows under Different Flow Regimes

Consider water flowing from a pool of water having a constant water level to a channel. At the entrance of the channel (i.e. at the beginning) the more the water moving down, the more it will try to move faster (i.e. accelerate) towards the downslope due to the effect of gravity force and acceleration. However, as it moves down, accumulated friction force (from the channel bed and banks) will try to retard the flow speed and, at one stage, the accumulated friction force will become equal to the gravity force on the water acting towards the flow direction. When both the flows equate each other (i.e. $\Sigma F = 0$), as per Newton's second law (i.e. $\Sigma F = ma$, where ΣF is sum of all the forces, m is the mass of the object and a is the acceleration), the acceleration of the object (on which forces are acting) has to be 'zero' (as m can't be zero for a particular object). When the acceleration is zero, water will start flowing with a constant velocity, which is termed as 'uniform velocity'. This velocity will be maintained until the flow is disrupted by any obstruction/external influence, in this case, as freefall starts at the end of the channel. Flows at the beginning and at the end of the channel can be termed as 'varied flow' due to the acceleration and change in channel condition, respectively. Uniform flow occurs in between where flow depth (and velocity) remains constant. For particular channel conditions, wherever the water depth remains (without any interference) for certain discharge is called the 'normal depth'.

The depth of the uniform flow (and velocity) will depend on the slope of the channel; for a mild slope channel (Figure 7.1), the flow is likely to be subcritical with the depth higher than the critical depth (i.e. normal depth line will be higher than the critical depth line). Critical depth for a rectangular channel can be easily calculated if the q for the channel is known. All the slopes which will create a subcritical flow is called a 'subcritical slope', i.e. bed slope, $S_0 < S_c$ (S_c is the critical slope). In the Figure 7.1 uniform flow zone is shown in between two varied flow zones.

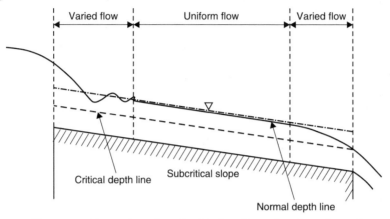

Figure 7.1. Formation of a subcritical uniform flow in a channel

Now for the same flow and source, if the downstream channel is made very steep (Figure 7.2), then the flow in the channel is likely to be turned into a supercritical flow (i.e. depth of flow lower than critical depth). In such a case, the normal depth line would be lower than the critical depth line for this particular flow. In this case, the uniform flow will occur until about the end of the channel where freefall commences. All the slopes which will create a supercritical flow is called a 'supercritical slope', i.e. bed slope, $S_0 > S_c$. In between, there is a slope (S_c) for which the channel flow will be critical and the critical depth line and normal depth line will merge into a single line (Figure 7.3).

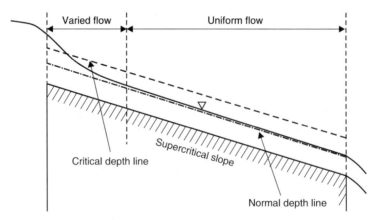

Figure 7.2. Formation of a supercritical uniform flow in a channel

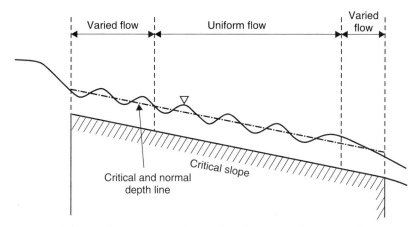

Figure 7.3. Formation of critical uniform flow in a channel

From Figure 7.3, it is clear that, unlike subcritical and supercritical flows, even within the uniform flow zone, the depth of water is not stable, but fluctuating. The reason for these fluctuations is that the critical flow is happening only at a particular point (a point on the tip of the energy curve, as shown in Figure 7.4) and to stay on that fine point for a longer period is difficult (as water gets disturbances from other sources, like wind, change of bottom/bank frictions, etc.). As such, the water level fluctuates up and down, as shown in Figure 7.4. This can be compared with a person trying to keep stable standing on a big ball. Figure 7.5 is a practical example of such critical flow on top of a weir, where, due to the formation of a critical flow, the water surface is not stable, but fluctuating.

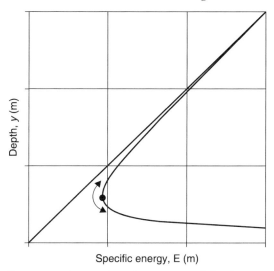

Figure 7.4. Fluctuations from a critical depth

Figure 7.5. Fluctuating critical flow on top of a weir

7.2 Uniform Flow Equation

For the derivation of the uniform flow equation, consider a longitudinal section of a rectangular open channel flowing under gravity, as shown in Figure 7.6. Other channel properties include:

Cross-sectional area: A
Wetted perimeter: P
Water depth: y
Longitudinal slope: S_0 (i.e. S_0 in 1)
Flow velocity: V
Hydraulics radius (R): A/P

Consider a block of water having a length of L and a depth of y, moving downwards in the channel (as shown in Figure 7.6) with a velocity of V. The shear stress on an object is proportional to its (velocity)^2, as such, shear stress on the water block = $K*V^2$, where K is a constant.

Shear Force (F) = Shear Stress*Area = $K*V^2*P*L$

Considering the weight of the block as W, the horizontal component of the weight = $W*\sin\theta = \gamma*A*L*\sin\theta$

For a very mild slope, $\sin\theta$ can be approximated as S_0. Therefore, the horizontal component of the weight force = $\gamma*A*L*S_0$. Summing all the forces (friction force and the horizontal component of the weight) yields

$$\gamma*A*L*S_0 - K*P*L*V^2 = 0$$

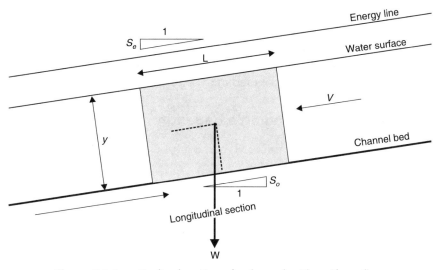

Figure 7.6. Longitudinal section of a channel with uniform flow

$\Rightarrow$ $$\gamma * A * L * S_0 = K * P * L * V^2$$

$\Rightarrow$ $$V^2 = \gamma * (A/P) * (1/K) * S_0$$

$\Rightarrow V^2 = C^2 * R * S_0$, where, $R = A/P$ (hydraulic radius) and $C^2 = \gamma/K$ (constant)

$\Rightarrow$ $$V = C\sqrt{R * S_0} \tag{7.1}$$

This equation was first derived by a French engineer, Antoine de Chezy and, as such, the equation is named as Chezy's equation and the constant C is called Chezy's constant. Later on, an Irish engineer, Robert Manning discovered that C is not really a constant, rather $C = R^{1/6}/n$, where n is a constant dependant on the channel surface roughness. Replacing this C value into the Equation 7.1 yields

$$V = (1/n)R^{2/3}S_0^{1/2} \tag{7.2}$$

This equation is called Manning's equation and the n is called Manning's roughness coefficient. For discharge/flow, the same equation can be written as:

$$Q = (1/n)AR^{2/3}S_0^{1/2} \tag{7.3}$$

where A is the cross-sectional area of the flow. This equation is widely-used all over the world for the calculations of flow in natural as well as man-made channels. Over the years, researchers evaluated different values of n for many different surfaces under different conditions. In a controlled section having any surface (i.e. material), for a particular flow if the hydraulic radius (R) and cross-sectional area (A) can be measured, then, using the above equation, n for the particular surface can be evaluated. Table 7.1 shows recommended n values for different surfaces under different conditions. More detailed n values for open natural and constructed channels are provided in the Appendix A.

7.3 Solutions of Manning's Equation

There are two types of solutions: 1) solving for Q and 2) solving for y.

Solving for 'Q'

If the channel slope, surface roughness (n), depth (y) and channel cross-sectional properties are known, then, from the channel geometric properties, A, P and R can be calculated. Therefore, Q can be calculated using Equation 7.3.

Table 7.1. Recommended n values for different surfaces

Surface material	Manning's n	Surface material	Manning's n
Asbestos cement	0.011	Galvanized iron	0.016
Asphalt	0.016	Glass	0.01
Brass	0.011	Gravel, firm	0.023
Brick	0.015	Lead	0.011
Canvas	0.012	Masonry	0.025
Cast-iron, new	0.012	Metal – corrugated	0.022
Clay tile	0.014	Plastic	0.009
Concrete – steel forms	0.011	Rubble masonry	0.017
Concrete (Cement) – finished	0.012	Steel – smooth	0.012
Concrete – wooden forms	0.015	Steel – new unlined	0.011
Copper	0.011	Steel – riveted	0.019
Corrugated metal	0.022	Wood – planed	0.012
Earth, smooth	0.018	Wood – unplaned	0.013

Source: https://www.engineeringtoolbox.com/mannings-roughness-d_799.html

Solving for 'y'

If the channel slope, surface roughness (n), channel cross-sectional properties and discharge (Q) are known, then the trial and error method has to be applied in order to evaluate the channel depth, y. The following steps must be followed for such a solution (the latest scientific calculators can solve this directly):

Step 1: Guess depth, y
Step 2: Calculate geometric properties, A, P & R
Step 3: Calculate discharge, Q
Step 4: Compare with the required value of Q
Step 5: Repeat steps 1 to 4
Step 6: Stop when the required value of Q is found

7.4 Details of Manning's Roughness

Factors affecting Manning's Roughness

- Surface Roughness
- Vegetation
- Channel Irregularity
- Channel Alignment
- Obstruction
- Size and Shape of Channel
- Stage and Discharge

For a particular channel section, by considering all the above-mentioned factors, an ultimate/integrated Manning's n value can be evaluated using the following equation (Cowan, 1956):

$$n = (n_0 + n_1 + n_2 + n_3 + n_4)*m_5$$

Where, n_0 is the basic value of n for a straight, smooth channel, n_1 is the value added for surface irregularities, n_2 is the value for variations in shape and size, n_3 is the value for obstruction, n_4 is the value for vegetation and m_5 is the correction factor for the meandering of the channel.

7.5 Compound Channel

In reality, a natural channel may have different roughness at different heights/levels, as shown in Figure 7.7, where the lower portion is concrete, whereas the upper (i.e. floodplain) portion is overgrown grass and shrubs.

Even for a complete natural channel, usually the roughness of the main channel is low as the roughness is mainly contributed from the bed and bank surfaces without large obstructions (i.e. large trees). In contrast, the roughness of a floodplain is high due to the existence of several large objects, including trees (Figure 7.8). Furthermore, water depth on floodplains is lower compared to the main channel water depth and for shallower water, the relative roughness is higher.

Figure 7.7. Photo of a typical compound channel

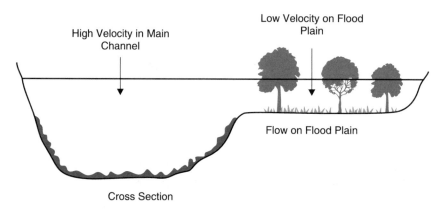

Figure 7.8. Typical cross-section of compound channel having floodplain

For the solutions of determining discharge from such compound channels, first the whole channel has to be subdivided based on channel geometry and/or roughness characteristics (Figure 7.9). Each subdivision will be named as '1, 2, 3 and so on'. Then, for each sub-channel, from the channel geometry, the parameters A, P and R will be calculated. Then, knowing the channel's longitudinal slope (S_0) and Manning's roughness coefficient, using the Manning's equation (Equation 7.3), discharge (Q) will be calculated for each sub-channel. Eventually, the final Q value would be the sum of all the individual Q values from all the sub-channels. It is to be noted here that, for the calculation of wetted

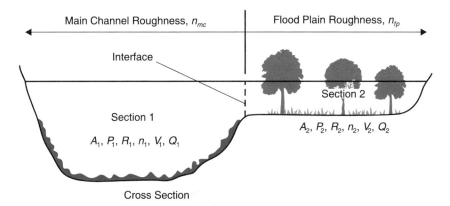

Figure 7.9. Subdivision of compound channels

perimeter (P) for each sub-channel, the contribution of interface (shown in Figure 7.9) has to be ignored. The contribution of P in the Manning's equation comes through the surface roughness and, as such, the longer the P, the higher the cumulative roughness is and, hence, the lower the velocity. However, the interface is not contributing to roughness as this is a virtual surface only. Nonetheless, for the calculation of A, the whole sub-section has to be considered. For such compound channels, if Q and other channel geometric properties are known, while y is unknown, then for the solution of y the process is not straight forward, rather, it requires trials (i.e. assuming y, then calculating Q for this particular y using the above-mentioned method and then comparing the calculated Q with the given Q. Then repeating the same process until the calculated Q value becomes equal or very close to the given Q value). Such trial method is usually lengthy and clumsy, however, this is made easy through the use of advanced scientific calculators or Excel spreadsheets.

7.6 Conveyance of Open Channel

In Manning's equation, for a particular channel section with a known surface property, all other parameters except the longitudinal slope (S_0) are constant. As such, the Manning's equation can be simplified as:

$$Q = \left(\frac{1}{n}\right)AR^{2/3}S_0^{1/2} = K * S_0^{1/2} \tag{7.4}$$

where $K = \left(\frac{1}{n}\right)AR^{2/3}$, is a constant for a particular channel section and known as 'channel conveyance' (Chow, 1973). Conveyance of a channel section with a particular surface property increases with the increase of the area (A) and the hydraulic radius (R), but decreases with the increase of the wetted perimeter. A channel section having the least wetted perimeter for a given area has the maximum conveyance. As such, a channel of a semicircular shape has the least perimeter among all the sections with the same area.

7.7 Design of Uniform Flow Channel

For the design of a channel, the basic aim is to convey the maximum amount of discharge through a minimum possible area. To achieve this, a channel section should generally be designed for the best hydraulic efficiency (i.e. highest channel conveyance), but should be modified for practicability. As such, although a semicircular shape has the highest conveyance for a given area, it is seldom used as a channel section due to its inconvenience for maintenance.

On the basis of channel construction method and material, there are two types of channels; a) Non-erodible (lined) channel and b) Erodible (natural) channel. Figures 7.10 and 7.11 show some typical photos of lined channels, whereas Figure 7.12 shows a photo of a natural channel.

The traditional urban practice was to construct lined channels, mainly in order to avoid channel erosion and to provide efficient drainage. However, due to rapid urbanisation and urban developments, many such efficient drainage systems (which discharge to a downstream point) became a cause of flash floods in urban areas, without providing enough time for the residents to evacuate. Moreover, in regard to

Figure 7.10. Typical rectangular shaped lined channel

Figure 7.11. Typical trapezoidal shaped lined channel

Figure 7.12. Typical natural channel with plants

stormwater pollutants, such efficient drainage systems are just shifting all the pollutants to a downstream point, without doing any treatment. As such, a recent recommendation from urban planners is to not provide such lined channels, but rather to construct natural channels with suitable plants which will stabilise the soil (i.e. will prevent erosion). These natural channels (with grass and plants) also help in trapping sediments as well as nutrients. However, these natural channels (with plants/grass) have a higher roughness, which will retard the flow velocity and hence for a specific discharge will cause higher flood levels.

The three major parameters for designing a uniform flow channel are:

(a) Size
(b) Shape
(c) Slope

Size is mainly governed by the 'design discharge' or the amount of flow to be conveyed. The shape of the channel depends on a few factors, such as the 'availability of land', 'ease of maintenance' and 'safety concern for people/animals'. To provide easy maintenance and avoid safety concerns, a trapezoidal shaped channel (Figure 7.11) is more common in an urban setup. Although, if one wants to convey more discharge within the occupied space, then a rectangular shaped channel (Figure 7.10) will convey more discharge within the same occupied landscape. However, in regard to regular maintenance, such shapes are not preferred due to the vertical sides. This shape also has some safety concerns, and, therefore, should be provided only as a last option, when other options are not suitable/achievable. If such a channel has to be constructed, then proper fencing must be provided along both the banks for safety. Moreover, proper steps down from the bank should be provided for maintenance access. In regard to the slope of the channel, the required slope can be calculated using Manning's equation for a known discharge and channel geometry with the criteria of intended velocity to be supercritical/subcritical (as critical flow is not stable). In general, supercritical flow is not preferred for reason of safety, as well as to minimise/avoid erosion. In the case of channel design, usually there are not many choices in regard to slope, as the channel slope mainly follows the natural topographic slope and for a mild topography channel the slope exactly follows the natural topographic slope. Otherwise, the channel will become too deep or exposed above ground surface. For a very flat topography this is an issue, as the channel should have a slope to convey water downstream by gravity. In this case, however, the ground surface is flat and at downstream of the channel it becomes too deep and, therefore, a safety concern (Figure 7.13). The same situation occurs when a mild channel passes through a mound or hilly areas, where the channel becomes very deep (Figure 7.14). For a steep topography, if the channel slope is kept mild, towards downstream

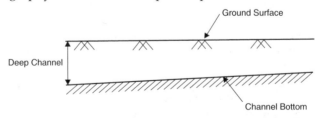

Figure 7.13. A channel running through a flat topography

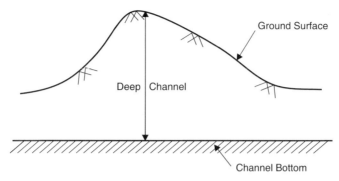

Figure 7.14. A mild channel running through a hilly topography

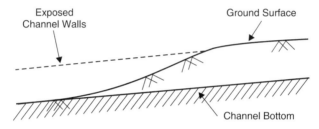

Figure 7.15. A mild channel running through a steep topography

the channel walls will be exposed above the ground surface (Figure 7.15). An alternate solution is to construct a meandering channel having a mild slope, which is similar to a road coming down from a steep hill (to avoid a steep unsafe road, it is constructed in a zig-zag fashion while keeping the slope relatively mild).

Worked Example 1

Water flows in the channel shown in the figure below under uniform flow conditions. Determine the discharge in the channel when the depth of water is 3.0 m. Channel longitudinal slope is 1 cm/m. Manning's n values for the relevant sections are shown in the figure.

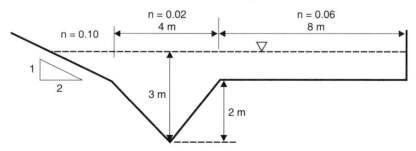

Solution:

The longitudinal slope is 1 cm/metre, which is 1 cm per 100 cm and equates to $1/100 = 0.01 = S_0$.

Based on geometry and channel roughness, the channel section should be divided into three sub-sections; the left portion is 'section 1', the middle portion is 'section 2' and the right portion is 'section 3'. Now, discharges need to be calculated separately for each section using, Manning's equation.

Section 1

Depth = $3 - 2 = 1$ m and width (with 1:2 ratio) = $2*1 = 2$ m

Area, $A = \frac{1}{2}*2*1 = 1$ m^2

Perimeter, $P = \sqrt{(1^2 + 2^2)} = 2.236$ m

$R = A/P = 0.447$

Using Manning's equation, $Q_1 = 1/n*A*R^{2/3}*\sqrt{S_0}$

$\qquad = 1/0.1*1*0.447^{2/3}*\sqrt{0.01} = 0.585$ m^3/s

Section 2

Area, A = Triangular area + rectangular area = $\frac{1}{2}*4*2 + 1*4 = 8$ m^2

Perimeter, $P = 2*\sqrt{(2^2 + 2^2)} = 5.66$ m (considering only surfaces on the triangular area)

$R = A/P = 1.413$ m

Using Manning's equation, $Q_2 = 1/n*A*R^{2/3}*\sqrt{S_0}$

$\qquad = 1/0.02*8*1.413^{2/3}*\sqrt{0.01} = 50.38$ m^3/s

Section 3

Area, $A = 8*1 = 8$ m^2

Perimeter, $P = 8 + 1 = 9.0$ m (bottom and right side surfaces)

$R = A/P = 0.889$ m

Using Manning's equation, $Q_3 = 1/n*A*R^{2/3}*\sqrt{S_0}$

$\qquad = 1/0.06*8*0.889^{2/3}*\sqrt{0.01} = 12.33$ m^3/s

Therefore, total discharge = $Q_1 + Q_2 + Q_3 = 63.3$ m^3/s

Worked Example 2

Water flows in the channel shown in the figure below under uniform flow conditions. Determine the discharge in the channel when the depth $y = 1.85$ m. The channel longitudinal slope is 15 mm/m. Manning's n values for the relevant sections are shown in the figure.

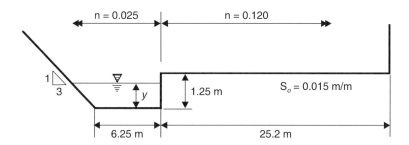

Solution:

The longitudinal slope is 15 mm/metre, which is 1.5 cm per 100 cm and equates to $1.5/100 = 0.015 = S_0$.

Based on geometry and channel roughness, the channel section should be divided into two sub-sections; the left portion is 'section 1' and the right portion is 'section 2'. Now, discharges need to be calculated separately for each section using, Manning's equation.

Section 1

Top width of the trapezoidal area = $6.25 + 1.85*3 = 11.80$ m

Bottom width of the trapezoidal area = 6.25 m

Depth = 1.85 m

Area, $A = \frac{1}{2}*(11.8 + 6.25)*1.85 = 16.7$ m^2

Perimeter, $P = 1.25 + 6.25 + \sqrt{\{1.85^2 + (3*1.85)^2\}} = 13.35$ m

$R = A/P = 1.25$ m

Using Manning's equation, $Q_1 = 1/n*A*R^{2/3}*\sqrt{S_0}$

$\qquad\qquad = 1/0.025*16.7*1.25^{2/3}*\sqrt{0.015} = 94.93$ m^3/s

Section 2

Area, $A = (1.85 - 1.25)*25.2 = 15.12$ m^2

Perimeter, $P = (1.85 - 1.25) + 25.2 = 25.8$ m (considering bottom and right sides only)

$R = A/P = 0.586$ m

Using Manning's equation, $Q_2 = 1/n*A*R^{2/3}*\sqrt{S_0}$

$\qquad\qquad = 1/0.12*15.12*0.586^{2/3}*\sqrt{0.015} = 10.81$ m^3/s

So, total discharge = $Q_1 + Q_2 = 105.74$ m^3/s

Worked Example 3

Water flows in the channel shown in the figure below under uniform flow conditions. Determine the discharge in the channel when the depth of water is 3.0 m. The channel slope is 15 m/km.

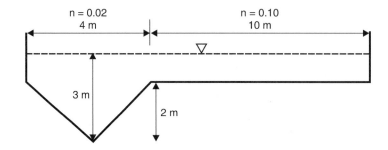

Solution:

The longitudinal slope is 15 m/km, which is 15 m per 1000 m and equates to $15/1000 = 0.015 = S_0$.

Based on geometry and channel roughness, the channel section should be divided into two sub-sections; the left portion is 'section 1' and the right portion is 'section 2'. Now, discharges need to be calculated separately for each section using, Manning's equation.

Section 1

Area, A = Triangular area + rectangular area = ½*4*2 + 1*4 = 8 m²

Perimeter, $P = 1 + 2*\sqrt{(2^2 + 2^2)} = 6.66$ m (considering surfaces on the triangular area and left side surface)

$R = A/P = 1.20$ m

Using Manning's equation, $Q_1 = 1/n*A*R^{2/3}*\sqrt{S_0}$

$\qquad = 1/0.02*8*1.20^{2/3}*\sqrt{0.015} = 55.32$ m³/s

Section 2

Area, $A = 10.0*1.0 = 10$ m²

Perimeter, $P = 10 + 1.0 = 11$ m (considering bottom and right sides only)

$R = A/P = 0.909$ m

Using Manning's equation, $Q_2 = 1/n*A*R^{2/3}*\sqrt{S_0}$

$\qquad = 1/0.10*10*0.909^{2/3}*\sqrt{0.015} = 11.49$ m³/s

Therefore, total discharge = $Q_1 + Q_2 = 66.81$ m³/s

Worked Example 4

Water flows in the channel shown in the figure below under uniform flow conditions. Determine the discharge in the channel when the depth of water is 3.0 m. The channel longitudinal slope is 10 m/km. Manning's 'n' values for the relevant sections are shown in the figure.

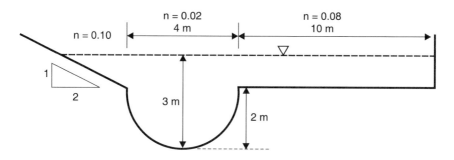

Solution:

The longitudinal slope is 10 m/km, which is 10 m per 1000 m and equates to $10/1000 = 0.01 = S_0$.

Based on geometry and channel roughness, the channel section should be divided into three sub-sections; the left portion is 'section 1', the middle portion is 'section 2' and the right portion is 'section 3'. Now, discharges need to be calculated separately for each section using, Manning's equation.

Section 1

Depth = 3 − 2= 1 m and width (with 1:2 ratio) = 2*1 = 2 m

Area, $A = \frac{1}{2}*2*1 = 1$ m^2

Perimeter, $P = \sqrt{(1^2 + 2^2)} = 2.236$ m

$R = A/P = 0.447$

Using Manning's equation, $Q_1 = 1/n*A*R^{2/3}*\sqrt{S_0}$
$$= 1/0.1*1*0.447^{2/3}*\sqrt{0.01} = 0.585 \text{ m}^3/\text{s}$$

Section 2

Area, A = semi-circular area + rectangular area = $\frac{1}{2}*\pi*2^2 + 1*4 = 10.28$ m^2

Perimeter, $P = 2*\pi*2/2 = 6.28$ m (considering only semi-circular surface)

$R = A/P = 1.637$ m

Using Manning's equation, $Q_2 = 1/n*A*R^{2/3}*\sqrt{S_0}$
$$= 1/0.02*10.28*1.637^{2/3}*\sqrt{0.01} = 71.39 \text{ m}^3/\text{s}$$

Section 3

Area, $A = 10*1 = 10$ m^2

Perimeter, $P = 10 + 1 = 11.0$ m (bottom and right side surfaces)

$R = A/P = 0.909$ m

Using Manning's equation, $Q_3 = 1/n*A*R^{2/3}*\sqrt{S_0}$
$$= 1/0.08*10*0.909^{2/3}*\sqrt{0.01} = 11.73 \text{ m}^3/\text{s}$$

Therefore, total discharge = $Q_1 + Q_2 + Q_3 = 83.71$ m^3/s

Worked Example 5

Water flows in a 3.0 m wide rectangular channel under uniform flow conditions. The Manning roughness for the channel is $n = 0.035$ and the channel has a bed slope of 1 in 200. The discharge was measured to be 6.5 m^3/s. Determine the depth of the flow in the channel. You can adopt a flow tolerance of ±5% in your calculations.

Solution:

Channel width, $B = 3.0$
Channel slope, $S_0 = 1/200 = 0.005$, $n = 0.035$ and $Q = 6.5$

Using Manning's equation, $Q = \left(\dfrac{1}{n}\right)AR^{2/3}S_0^{1/2}$

$$6.5 = \frac{1}{0.005} * (3 * y) * \left(\frac{3 * y}{3 + 2y}\right)^{2/3} * 0.005^{1/2}$$

$$6.5 = 6.061 * y * \left(\frac{3 * y}{3 + 2y}\right)^{2/3}$$

Through trial, for $y = 1.35$, $Q = 6.52$.

Therefore, final answer is 1.35 m.

Worked Example 6

Water flows in a triangular channel (side slope 1H:2V) under uniform flow conditions as shown in the figure along side. The Manning roughness for the channel is $n = 0.02$ and the channel has a bed slope of 1 in 200. The discharge was measured to be 6.0 m^3/s. Determine the depth of the flow in the channel.

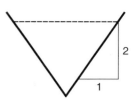

Solution:

Channel slope, $S_0 = 1/200 = 0.005$, $n = 0.02$ and $Q = 6.0$

Let depth of water in the channel be y. Width of the flow for a depth of y is also y.

Area, $A = \frac{1}{2} * y * y = y^2/2$

Perimeter, $P = 2 * \sqrt{((1/2)^2 + 1^2)} * y = 2 * \sqrt{5}/2 * y = \sqrt{5} * y$

$R = A/P = 0.2236 * y$

Using Manning's equation, $Q = \dfrac{1}{n} A R^{2/3} S_0^{1/2}$

$$6.0 = \frac{1}{0.02} * (y^2/2) * (0.2236 * y)^{2/3} * 0.005^{1/2}$$

$$6.0 = 0.651 * (y)^{8/3}$$

$$(y)^{8/3} = 9.22$$

$$(y) = (9.22)^{3/8} = 2.30 \text{ m}$$

Worked Example 7

Water flows in a channel (having a crown-shaped middle portion, as shown in the figure below, with sides' slope of (1H:2V) under uniform flow conditions. The Manning's roughness for the channel is $n = 0.025$ and the channel has a bed slope of 1 in 200. The discharge was measured to be 7.0 m^3/s. Determine the depth of the flow in the channel.

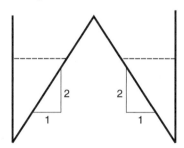

Solution:

Channel slope, $S_0 = 1/200 = 0.005$, $n = 0.025$ and $Q = 7.0$

As the two sections are identical, we can calculate discharge for one section and then multiply by '2' to get the total discharge.

Let depth of water in the channel be y. Width of the flow for a depth of y is $y/2$.

Area, $A = \frac{1}{2} * (y/2) * y = y^2/4$

Perimeter, $P = y + \sqrt{((1/2)^2 + 1^2)} * y = y + \sqrt{5}/2 * y = 2.118 * y$

$R = A/P = 0.118 * y$

Using Manning's equation, $Q = \dfrac{1}{n} A R^{2/3} S_0^{1/2}$

$$7.0 = 2 * \frac{1}{0.025} * (y^2/4) * (0.118 * y)^{2/3} * 0.005^{1/2}$$

$$7.0 = 0.34 * (y)^{8/3}$$

$$(y)^{8/3} = 20.59$$
$$(y) = 20.59^{3/8} = 3.11 \text{ m}$$

Worked Example 8

In a flat topography, a drainage channel is flowing with a bed slope of 1 in 500 (mild slope). At the upstream side, the depth of the channel is 2 m from the ground surface. If the ground surface remains the same, what would be the depth of the channel at a distance 5 km towards downstream?

Solution:

A slope of 1 in 500 means, a drop of 1 m for a horizontal distance of 500 m. So, for a distance of 5 km (i.e. 5000 m) the drop will be 5000/500 = 10 m.

So, the depth of the channel at a distance 5 km towards downstream would be = 10+2 = 12 m.

Worked Example 9

For a steep topography, having a slope of 1 in 50, a drainage channel is carrying water downstream. The drainage channel has a slope of 1 in 200 and, at an upstream point, the depth of the channel from the ground surface is 2 m. If, at an upstream, the channel depth is 2 m from the ground surface, what would be the difference of levels between ground surface and the channel bottom at a distance 1 km downstream?

Solution:

A slope of 1 in 50 means, a drop of 1 m for a horizontal distance of 50 m. So, for a distance of 1 km (i.e. 1000 m) the ground surface will drop by 1000/50 = 20 m. Also, the channel will drop by 1000/200 = 5 m.

So, if the level of the ground at the upstream point is considered as '0'.

Level of the channel bottom = 0 – 2 = –2 m.

Level of the ground at 1 km downstream = 0 – 20 = –20 m.

Level of the channel bottom at 1 km downstream = –2 – 5 = –7 m.

Therefore, at 1 km downstream the channel bottom will be 20 – 7 = 13 m higher than the ground surface.

References

Chow, V.T. (1973). Open Channel Hydraulics. McGraw Hill Company Japan, ISBN 007 085906X.

Cowan, W.L. (1956). Estimating Hydraulic Roughness Coefficients: Agricultural Engineering 37(7): 473–475.

Hydraulic Modelling

8.1 Introduction

Chapter 5 discussed hydrologic modelling, which primarily involves calculating/predicting discharges from a certain catchment under a certain rainfall pattern/intensity. Hydrologic modelling mentions the amount of expected discharge(s) at the end of a catchment outlet, however, it is unable to mention the depth of flooding in a particular channel section or the surrounding topography. In regard to real flood visualization/concern, it is very important to know the depth of flow/flood in a particular channel section/locality, as the general community does not have a real perception on the level of danger from a particular amount of flow/discharge. Hydraulic modelling converts a certain amount of discharge into corresponding water depth based on channel geometry at a particular section. As such, hydraulic modelling is more important in regards to flood warning and/or determination of expected damage. There are two types of hydraulic modeling: Steady flow and unsteady flow. Unsteady flow modelling requires solving complex unsteady flow equations (continuity and momentum equations) under both spatial and temporal domains, making it a difficult and awkward process that calls for special expertise in order to achieve the solutions. Steady flow modelling, on the other hand, can be performed using Manning's equation and energy equations in two consecutive sections. Although real flow is unsteady, if steady flow analysis is done with the peak flows only as a worst case scenario, then the purpose of flood warning can be achieved. Nonetheless, such steady flow analysis with only peak flows will not be accurate for a complex river/creek network having long rivers/creeks in a large catchment.

8.2 Solution Process

Hydraulic model calculations for steady flows are performed using the energy equation and Manning's equation. In the previous chapters, the energy equation was considered without any energy loss. For the hydraulic modelling calculations, energy equation is considered with a loss, which is obvious for a natural channel having significant bed roughness/friction. Consider a natural river long-section, flowing from upstream (Section 2) to downstream (Section 1), as shown in Figure 8.1.

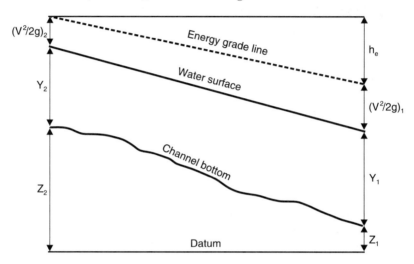

Figure 8.1. Typical river long-section with energy line (*Source*: USACE, 2010a)

As per the figure, the total energy at 'section 2' from the shown datum $= Z_2 + Y_2 + (V^2/2g)_2$ and the total energy at 'section 1' from the same datum $= Z_1 + Y_1 + (V^2/2g)_1$, where Z represents elevation to the channel bottom from the datum (i.e. elevation head), Y represents water surface height from the channel bottom (i.e. pressure head) and V represents the flow velocity in the respective section. Energy equations in the two sections can be equated using an energy loss, h_e. Therefore, the conservation equation becomes,

$$Z_2 + Y_2 + (V^2/2g)_2 = Z_1 + Y_1 + (V^2/2g)_1 + h_e \qquad (8.1)$$

The energy loss (h_e) can be approximated as $S_e * L$, where S_e is the slope of the energy line and L is the length of the channel section. S_e can be calculated using Manning's equation (Equation 8.2), provided discharge (Q), water depth (y), channel cross-sectional properties and channel roughness (n) are known.

$$Q = (1/n)AR^{2/3}S_e^{1/2} \qquad (8.2)$$

Equation 8.1 is the basic equation for steady flow hydraulic modelling. However, from one equation, only one unknown can be solved, so all the parameters in Equation 8.1 have to be known, except one. Typically, Z values are known from land survey. V values are calculated from known/ assumed Y values using the given Q value and channel cross-sectional properties (gathered through channel survey). Using Manning's equation S_e and subsequently h_e can be calculated. Having all these data mentioned here, if one of the Y values is known, then the other can be calculated using Equation 8.1. However, the solution of the Y value using the Equation 8.1 is not straightforward, it requires trials. Suppose all the parameters in the right hand side (RHS) of Equation 8.1 are known, and Y_2 needs to be calculated. However, value of V in 'section 2' also depends on Y_2. As such, once the RHS of the equation is known, then the next step is to assume a Y_2, then calculate V at 'section 2' using given Q and assumed Y_2. Then the left hand side (LHS) parameters of the equation have to be added and compared with the RHS parameters' added value. If both the RHS and LHS values are same (or very close), then the assumed Y_2 is correct. Otherwise, the same procedure has to be followed with another assumed Y_2 until the two values (RHS and LHS) become almost the same. In some cases, even though both the Y values are unknown, one Y can be evaluated if it is at the location of any control section (i.e. weir, sluice gate, critical flow section) and the other Y can be calculated using the above-mentioned procedure. Then, considering this calculated Y value (i.e. Y_2) as the known value, Y value of further upstream section can be calculated using the same process. This procedure is continued until the evaluation of the Y value at a desired upstream distance. The overall solution process can be summarised as follows:

- Y_1 is either given or evaluated based on a control section criterion
- From Y_1, A and R values are calculated from the known cross-section data
- From the known Q, n, A and R values, S_e is calculated using Manning's equation
- From the calculated S_e value and known length (L) between the sections, h_e is calculated
- Y_2 is assumed, then V_2 is calculated using known Q and cross-section data
- LHS of the Equation 8.1 is matched with the calculated RHS of the equation
- The process is repeated until a good assumption of Y_2 is reached
- Considering the recent Y_2 value as the new Y_1, the new Y_2 value is calculated for further upstream section and the same process is continued up to the desired distance.

8.3 Data Requirements

Based on the procedure mentioned in the previous section, the solution process using Equation 8.1 requires the following data:

- Upstream discharge
- River/channel cross-sections (both upstream and downstream)
- Downstream control(s) or known water level
- Channel roughness

Upstream discharge is evaluated through catchment analysis and/or hydrologic modeling discussed in Chapter 5. In some cases, in the absence of proper discharge data/calculation, discharge is evaluated from the earlier established rating curve (discussed in Chapter 2) for that particular location (i.e. upstream). Channel cross-section data is established through the survey of channel banks and bed. Usually, a cross-section is presented as levels (reference levels) of a cross-section at different distances from a particular survey station (i.e. reference point). Downstream controls are implemented through either a known water level, critical depth, normal depth (calculated using bed slope, channel roughness and Manning's equation) or a rating curve (which converts discharge to depth at that particular location). The channel roughness value is taken (or derived) from the list of channel roughness values discussed in Chapter 7.

8.4 Hydraulic Modelling using HEC-RAS

HEC-RAS stands for Hydrologic Engineering Center's River Analysis System and is the most widely-used hydraulic model for one-dimensional analysis. Hydrologic Engineering Center is a part of the US Army Corps of Engineers (http://www.hec.usace.army.mil) who have developed this software. Although this chapter only describes the steady flow analysis, in fact, HEC-RAS also performs unsteady flow analysis. In addition, HEC-RAS is capable of performing some more sophisticated analyses, such as sediment flow calculations, flow calculations through a bridge, culvert, sluice gate, drop structure, weir and spillways. These analyses are not discussed in this chapter. Details of the software functionalities can be obtained from the HEC-RAS Users Manual (USACE, 2010b).

HEC-RAS operates from a central file named 'Project File', which contains names and locations of three associated files: 'Plan File', 'Geometry File' and 'Flow File'. Figure 8.2 shows the HEC-RAS main menu with provisions for descriptions and locations of the 'Project File' and other associated files. Figure 8.3 shows the HEC-RAS menu option to edit geometry and flow data files. These files can also be opened/edited from the icons next to the 'save' icon. Editing 'Geometry File' opens a new

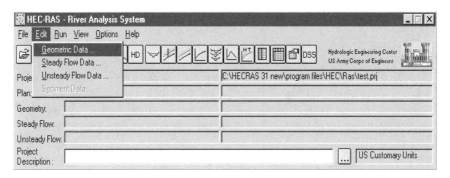

Figure 8.2. HEC-RAS main menu/window (*Source*: USACE, 2010b)

Figure 8.3. Editing Geometry/Flow data in HEC-RAS (*Source*: USACE, 2010b)

window, as shown in Figure 8.4. In this window, the 'river reach' icon allows the user to draw a plan view of a river/channel network (while drawing such a network, by default the first point is considered upstream and the last point is downstream).

By using the 'cross-section' button at the top-left corner, cross-section data is inserted for each cross-section (in the Figure 8.4 number, location and width of the traverse lines are the number, relative position and relative width of inserted cross-sections). For each cross-section, a separate window will be opened, as shown in Figure 8.5. All the other cross-sections' data can be navigated from this window by pressing the arrow (up or down) button at the top-right corner. The 'cross-section' window contains elevations of several points (across the cross-section) at different distances, referred to as 'Station' (measured from a reference point). Real cross-sections can be viewed at the right side, as shown in Figure 8.6, where each dot point represents inserted data (one row) in the cross-section data. The 'cross-section' window also contains distances of downstream cross-section, left and right bank stations, Manning's roughness values at the channel and banks and the expansion/contraction co-efficient (if any).

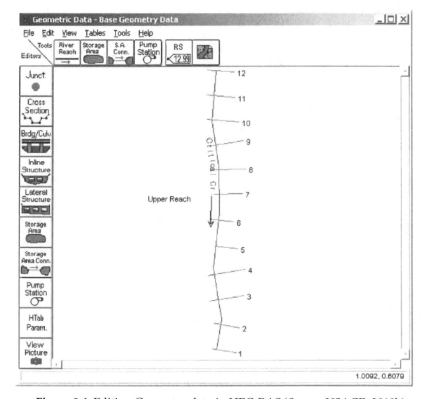

Figure 8.4. Editing Geometry data in HEC-RAS (*Source*: USACE, 2010b)

In the cross-section window, 'LOB' and 'ROB' stand for 'left overbank' and 'right overbank' respectively. In the USA, 'overbank' is the term used for floodplain, used in many other countries including Australia.

The 'steady flow data' window allows upstream flow data to be inserted. For each upstream location, a set of flow data can be inserted. Set of flows comprise different flow values under different conditions (i.e. 5, 10, 20, 50, 100 years ARIs). Figure 8.7 shows a complete flow data provided at four inflow locations in a river system with three flow data (10 yr, 50 yr and 100 yr).

Providing the known condition at the downstream end is termed as providing the 'boundary condition'. Figure 8.7 shows the window to provide boundary conditions. Four types of boundary conditions can be selected; 1) known water surface, 2) critical depth, 3) normal depth and 4) rating curve. Selecting 'normal depth' will require the bottom/bed slope at the location to be provided. And selecting 'rating curve' will require a set of discharge-water level data at the location to be provided.

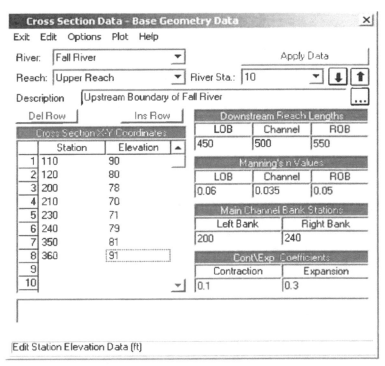

Figure 8.5. Editing cross-section data in HEC-RAS
(*Source*: USACE, 2010b)

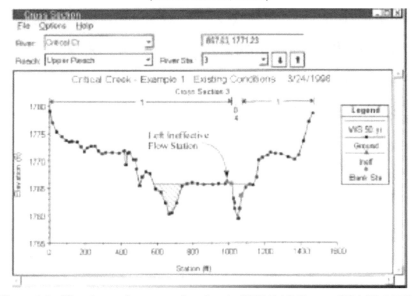

Figure 8.6a. Viewing real cross-section data in HEC-RAS (*Source*: USACE, 2010b)

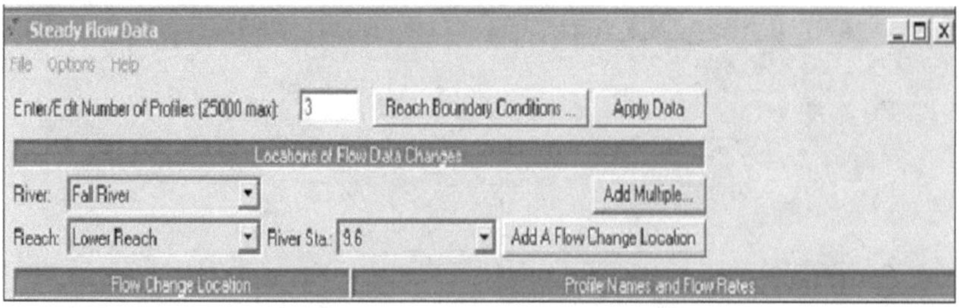

Figure 8.6b. Viewing real cross-section data in HEC-RAS (*Source*: USACE, 2010b)

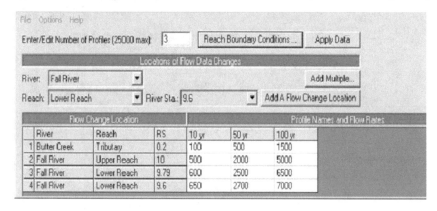

Figure 8.6c. Viewing real cross-section data in HEC-RAS
(*Source*: USACE, 2010b)

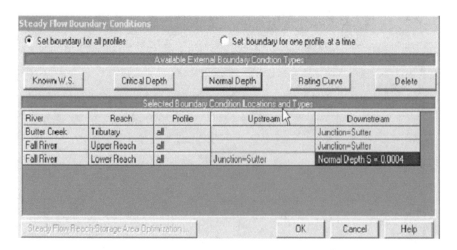

Figure 8.7. Providing boundary condition data in HEC-RAS
(*Source*: USACE, 2010b)

References

USACE (2010a). HEC-RAS River Analysis System Reference Manual for Version 4.1. US Army Corps of Engineers, Davis, CA. http://www.hec.usace.army.mil

USACE (2010b). HEC-RAS River Analysis System User's Manual for Version 4.1. US Army Corps of Engineers, Davis, CA. http://www.hec.usace.army.mil

Water Supply Systems

9.1 Introduction

Traditionally, urban water supply used to only deal with potable water, as it is safe to drink, pleasant to taste and suitable for other domestic purposes. In many urban areas (i.e. cities in USA, Canada, Australia), where single house living is predominating, more than 50% of water is being used for non-potable purposes (i.e. irrigation, toilet flushing and car washing etc.). The traditional practice was to supply the same potable water of high standard for all the purposes, as it was thought that to supply separate waters for potable and non-potable purposes would not be cost effective. On one hand, due to ever-increasing population growth and industrial processes, the demand for water has increased significantly. On the other side, due to the impacts of climate change, the availability of fresh water has become unreliable in many parts of the world, causing the water supply to be crucial in many cities. To overcome such acute situations, contemporary urban water authorities are supplying recycled water for non-potable purposes with the aim of reducing demand on potable water.

9.2 Water Consumption Pattern

Table 9.1 shows average per household and per capita water consumptions for different Australian states, based on Australian Bureau of Statistics' Water Account data for 2015-16 (ABS, 2017). It is to be noted here that these consumptions do not include industrial water demand, which was almost 7.5 times more than the household water demand for the whole of Australia. From the table, it is clear that household water demand is much higher for the states that have predominantly warm weather, as well as higher in-house irrigation practice. Table 9.2 shows the distributions of household water uses for different states of Australia (except Tasmania and Northern Territory) based on the Water Account data for 2000-1 by

Australian Bureau of Statistics (ABS, 2004). From the table, it is found that, for the two major states (New South Wales and Victoria), almost 50% of water is used for outdoor and toilet purposes. This proportion is even more than 60% for the states where in-house irrigation practices are predominant.

Table 9.1. Water consumptions in Australian cities

	Per capita consumption (L/person/day)	*Household consumption (L/household/day)*
New South Wales	199	533
Victoria	173	455
Queensland	212	549
South Australia	218	531
Western Australia	350	897
Tasmania	201	484
Northern Territory	434	1360
ACT	216	557
Australia	216	563

Table 9.2. Water consumptions in Australian cities

	Bathroom	*Toilet*	*Laundry*	*Kitchen*	*Outdoor*
NSW	26%	23%	16%	10%	25%
Victoria	26%	19%	15%	5%	35%
Qld	19%	12%	10%	9%	50%
SA	15%	13%	13%	10%	50%
WA	17%	11%	14%	8%	50%
ACT	16%	14%	10%	5%	55%
Australia	20%	15%	13%	8%	44%

NSW: New South Wales, Qld: Queensland, SA: South Australia, WA: Western Australia, ACT: Australian Capital Territory

In general, water consumptions vary with the climate, population and time during the day. Within a 24-hour period, less water is consumed during night time, while more water is consumed during early morning and evening. In regard to climate, more water is consumed in summer months, while less water is consumed in winter months. Figure 9.1 shows a typical variation of water consumption within a year (in Australia, June ~ August are the winter months, while December ~ February are the summer months). Figure 9.2 shows the typical diurnal variations of water consumptions.

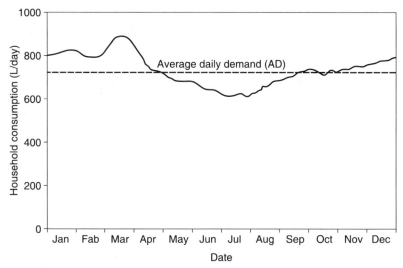

Figure 9.1. Typical annual variation of Australian water demand

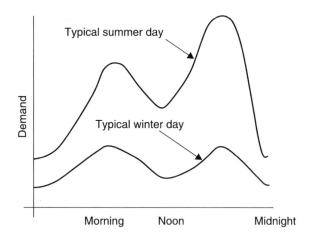

Figure 9.2. Typical diurnal variations of daily water consumption

Parameters used for Water Consumption Calculations

The following parameters are generally used for the calculations of water consumption and demand.

Average Day Demand (AD): Is the annual water consumption by a city/locality divided by the number of days in the recorded year.

Mean Day Maximum Month Demand (MDMM): Is the maximum monthly demand in a city/locality in a year, then divided by the days in that month.

Maximum Day Demand (MD): Is the maximum consumption recorded in a day for a city/locality.

Maximum Hour Demand (MH): Is the maximum consumption recorded in an hour for a city/locality.

Based on numerous data, traditionally, for the purpose of design, 'MD' and 'MH' are calculated using the following relationships:

MD = PDF × AD, where 'PDF' is the Peak Day Factor and assumed as:
PDF = 1.5 for population > 10,000
PDF = 2 for populations < 2,000

MH = PHF × MD/24, where 'PHF' is the Peak Hour Factor and assumed as:
PHF = 2 for population > 10,000
PHF = 5 for populations < 2,000

In the above-mentioned relationships, 'PDF' and 'PHF' for the intermediate populations (between 2,000 and 10,000) need to be calculated through interpolation between the specified values.

9.3 Estimation of Demand

For the planning and design of a water supply system for a city/locality, the very first step is the estimation of water demand. This demand is not the current demand, rather an estimated demand for a future population based on the design life of the system. As the population growth for a locality is affected by many factors, such as weather, accessibility to the amenities, accessibility to the city centre, affordability, economic activities and many others, it is difficult to accurately estimate the future population. However, for the purpose of a rough design, there are few commonly used models for the estimations of future population. Nonetheless, adoption of any of these models requires justification based on population data from the past years, which is gathered from the census data for a city/country. The following are the few commonly-used population growth models:

Arithmetic Growth Model: This model considers that the rate of the growth of the population is constant (per year), which can be expressed as:

$$\frac{dP}{dt} = C$$

Which yields (through integration),

$$P_t = P_0 + Ct \qquad (9.1)$$

where P_t is the population after t years, P_0 is the initial population at any base year (from when estimation is commenced) and C is the growth rate (usually per year). Figure 9.3 shows a graphical growth pattern for 40 years for a population having an initial population of '1000' and the growth rate of 100 per year.

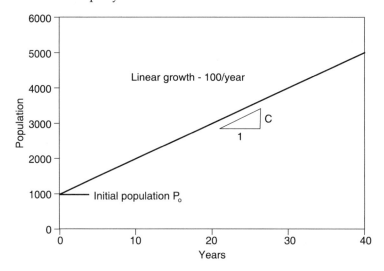

Figure 9.3. Graphical representation of arithmetic growth model estimation

Exponential Growth Model: This hypothesis was first proposed by Thomas Malthus in 1798, who claimed that the population growth is not arithmetic (or linear), but exponential, which is similar to the growth rate of bacteria and some insects. According to Thomas, food production is arithmetic, whereas population growth is exponential, as such cannot be matched with increased food production, unless population growth is controlled. According to his theory, population growth is proportional to the current population, i.e. $\frac{dP}{dt} = CP$ which yields (through integration),

$$P_t = P_0 * e^{Ct} \qquad (9.2)$$

where C is the exponential growth rate constant. For an initial population of '1000', having an exponential growth rate of 10% (i.e. $C=0.1$), the population after 40 years would be '54,598'. Figure 9.4 shows the graphical representation of the exponential growth pattern, in conjunction with the arithmetic growth pattern mentioned earlier.

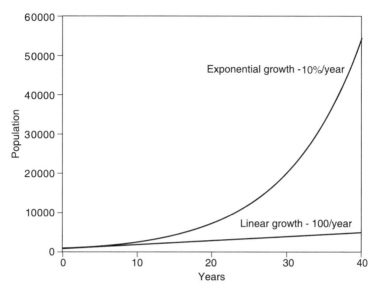

Figure 9.4. Comparison of arithmetic and exponential growth models

Logistic Growth Model: From Figure 9.4, it is clear that for a long period estimation, the exponential growth model provides a very high estimation, which is not realistic. As such, a third method was developed, which considers that population growth is not unlimited, rather it is limited by a carrying capacity and this carrying capacity is the upper limit to population growth. This method proposes a modified exponential constant incorporating 'maximum carrying capacity' (of a city/locality) as follows:

$$C' = C_{max}\left(1 - \frac{P}{K}\right)$$
(9.3)

where C' is the exponential growth rate constant, P is the population at any time, K is the maximum carrying capacity and C_{max} is the maximum growth rate. When, P becomes equal to K, $C' = 0$.

The final growth model is:

$$\frac{dP}{dt} = C_{max}\left(1 - \frac{P}{K}\right) * P \text{ which yields (through integration),}$$

$$P_t = \frac{K}{1 + \left[\left(\frac{K - P_0}{P_0}\right)e^{-C_{max}t}\right]}$$
(9.4)

For an initial population of '1000' and a maximum exponential growth constant of 10%, Figure 9.5 shows a representation of the logistic growth pattern for a period of 40 years with a maximum carrying capacity of '5000'. It is to be noted here, that this growth pattern is significantly affected by the 'carrying capacity'. In this example, if the maximum carrying capacity were '10,000', then the graph would have been quite different. Figure 9.6 shows the comparison of estimations through three

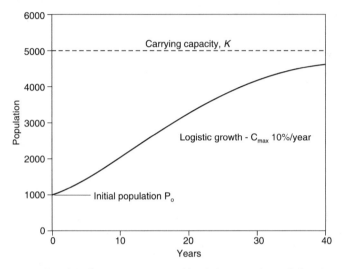

Figure 9.5. Graphical representation of logistic growth model estimation

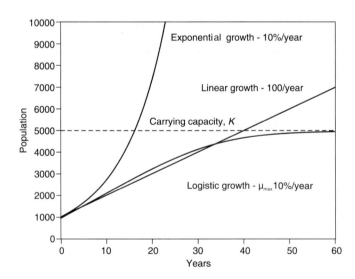

Figure 9.6. Comparison of estimations through three different models

different models for an estimation period of 60 years. From the figure, it is clear that the exponential growth model's estimation is far away from the other two estimations, within 23 years' time the estimation exceeds the upper boundary of the graph (i.e. 10,000). If a period of 100 years is considered for such an estimation (Figure 9.7), the exponential model will estimate a population of more than 22 million with a same growth rate (i.e. 10%) and the estimations through the two other methods are not traceable. As such, it can be concluded that the exponential growth model is not realistic for a long-term estimation. It is recommended that such estimation should be performed for relatively shorter periods (i.e. 20 years) and before adopting any method, the method should be tested with the past population data.

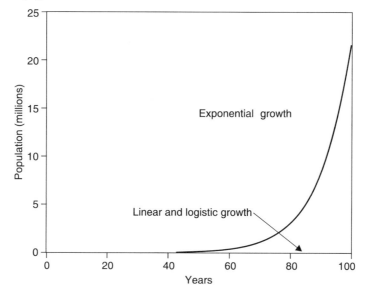

Figure 9.7. Comparison of estimations for a period of 100 years

9.4 Water Supply System Components

Once population is estimated, then the total daily demand of water is calculated using average/maximum daily demand. Then, based on this total demand, different components of the water supply system are designed. The following are the major components of a typical water supply system:

- Raw Water Source
- Water Treatment Plant
- Trunk Mains
- Distribution Reservoirs

• Water Distribution Pipe Network (Pipes, Valves, Pumps, Hydrants)

Among raw water sources, dams/reservoirs, groundwater aquifers and rivers are the main resources widely used for extraction of water. Selection of a particular source mainly depends on availability, feasibility, quality of water, quantity of water and/or cost-effectiveness of a specific source.

In regions where moderate to heavy rainfall occurs, a dam/reservoir is constructed across the river in order to hold rainwater and then to treat and supply the water to the residents/households. Although the construction of such dams have severe environmental/ecological consequences, numerous dams have been constructed around the world in order to supply water to the residents or for irrigation due to their ease of construction. For this raw source, water quality is significantly affected by the upstream catchment use, as such, authorities should take extra care on land uses of contributing catchments and prevention of some natural hazards, such as bushfire, erosion and pollutant spill. In addition, dams also have an impact on downstream river health. In particular, the ecological balance of downstream rivers suffer a lot as some aquatic species need a natural supply of water without such obstruction (and controlled supply). Authorities should monitor the water quality at the location and level of the intake (to treatment plant). Since during summer, stratification occurs within the lake, which hinders natural oxygenation of the lower portion of the dam/reservoir, the level of water intake is important. In Australia, most potable water is supplied from such dams/reservoirs.

Another common raw water source is a groundwater aquifer, and, in general, groundwater requires less treatment as it is naturally treated by natural filtration through the soil media. However, not all regions have this water source. It requires higher energy for pumping the water from the ground depending on the depth of the water table. Recently, in many regions, groundwater is also contaminated by different salts and/or chemical compounds like arsenic.

Another common source of water are rivers, provided they enough and continuous flow. Throughout history, many cities around the world have depended on river water. However, with the passage of time and over-extraction, many downstream regions of those rivers suffer from lack of water flow, resulting in conflicts among the downstream and upstream users/authorities. Also, due to the lack of proper governance and ever-increasing urbanisation, industrialisation and illegal dumping, the quality of many rivers in becoming worse and worse. This puts tremendous pressure on the existing water treatment plants, as many of those were not designed to treat a large amount of common contaminants and/or some uncommon contaminants.

Many arid regions do not have an enough of any of the above-mentioned sources which can be used for supply to the residents. Moreover, with the impacts of climate change, many countries are suffering from the lack of rain (i.e. drought). In such cases, the wealthy nations are extracting sea water through expensive desalination plants.

After extraction from the raw source, water requires treatment before it can be supplied to the residents. The level of treatment depends on the quality of the water in the raw source and may vary widely. After proper treatment, the treated water is conveyed through trunk mains (large water supply pipes) to the distribution reservoirs or nodes where smaller pipes carry it to the smaller areas. It is to be noted that, from the trunk mains, no individual connection is taken to the individual household. Figure 9.8 shows a photo of typical trunk mains supplying potable water.

Figure 9.8. Photo of trunk mains carrying water

The water supply system may or may not have a distribution reservoir, depending on the topography/altitude of the locality where water will be supplied. In some cases, in order to maintain a smooth supply of water without using energy, water is pumped into an overhead tank from where the water is supplied to the household through gravity. These overhead tanks are called as distribution/storage reservoirs. From storage reservoirs or trunk mains, water is supplied through a water distribution pipe network, which comprises of pipes (size decreasing as it goes to the end users), valves, pumps and/or rising mains. Storage reservoirs provide storage in order to meet daily fluctuations in use. They also provide storage for emergency firefighting use. They help stabilise pressures in the distribution system and are generally located at strategic points throughout a network (i.e. generally on a hill or high altitude).

Water distribution pipe network is the pipe network along the streets that supplies to individual consumers that are generally interconnected. Preferred option of such supply is that water should flow by gravity if supplied from the storage reservoir. Alternatively, if it is supplied without a storage reservoir, then pressure management along the pipe is an issue. Pumps and rising mains are used to boost the pressure at some downstream locations where the pressure drops. On the other hand, immediately downstream of such booster pumps, water pressure is

Figure 9.9. Photo of overhead storage tank

usually high, which causes bursting/leakage of such pipes. It is often difficult to maintain an optimum pressure throughout the pipe network. Figure 9.9 shows a photo of overhead storage tank.

9.5 Storage Tank Sizing

Storage tanks are designed to store and supply the intended daily maximum demand of a locality/community, as such, they should be big enough to handle the maximum consumption. Water is pumped into the storage tank and a constant rate of pumping is recommended in order to avoid additional management requirements of variable pumping rates. An average constant pumping rate (per hour) is determined by dividing the total daily demand by 24. In regard to the tank size, relevant authorities would like to provide an optimum size in order to minimise the cost. As such an optimum design is necessary with the aim that a storage volume should be just large enough to equalise pumping rate and consumption over maximum day demand. For the sizing calculation of overhead tanks, there are few commonly-used methods. To explain the methods, a typical hourly water demand for a community is presented in Figure 9.10 (corresponding values are shown in Table 9.3). An overhead tank size has to be determined for this consumption/demand. The first task is to determine the average (i.e. constant) pumping rate. For this particular example, the total consumption in 24 hours is 10,165 kL. Therefore, average pumping rate = 10,164.8/24 = 423.533 kL/hr. Cumulative pumping volumes (to the tank) and cumulative deviations of the consumptions from the pumped volumes are shown in Figure 9.11.

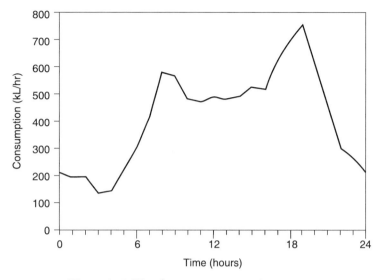

Figure 9.10. Hourly water consumption pattern

Table 9.3. Hourly water consumption values

Hour	Consumption (kL/hr)	Hour	Consumption (kL/hr)
1	196.7	13	483.7
2	196.7	14	492.8
3	136.3	15	529.1
4	144	16	522.3
5	227.1	17	622.3
6	302	18	697.2
7	415.6	19	756.2
8	583.6	20	606.4
9	567.8	21	454.2
10	486	22	302
11	472.4	23	265.7
12	492.8	24	211.9

Figure 9.11 shows the hourly consumptions along with the constant pumping rate and the hourly deviations from the pumping rate. From the figure, during the beginning hours (midnight onward), as the pumping is higher than the consumption, the reservoir will be filling with surplus

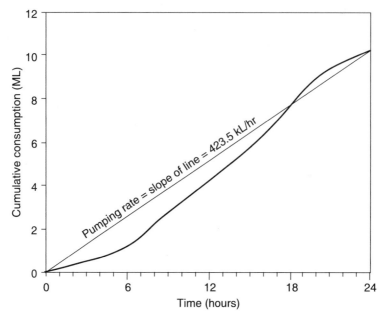

Figure 9.11. Cumulative consumptions and pumping volumes

water. Then, from early morning onward, as the consumptions are higher than the pumping rate, the reservoir will be emptying. Again at night, as consumptions reduce, the reservoir will start filling. Mathematically, the area of the curve above the constant pumping line is equal to the area of the curve below the constant pumping line. As such, if the total area of the curve around the pumping line can be calculated, then storage size required is "total area of the curve" divided by 2. Alternatively, the storage size required is the curve having positive (i.e. above the pumping rate) or negative (i.e. below the pumping rate) values.

Sizing Method 1

- Calculate the hourly differences between the pumping rate (inflow) and hourly consumptions
- Add all the negative values and positive values separately
- The required size is either the sum of negative or positive values

For the data provided in Table 9.3, the values of 'inflow-consumptions' are calculated, as shown in Table 9.4.

The sum of all the positive values and negative values are 1837.33 kL and 1837.34 kL, respectively (the slight difference is due to rounding values). Therefore, the required storage size is 1837.4 kL. In reality, the designer should round up this value to a higher rounded up value (i.e. 1840, 1850 or 1900 kL).

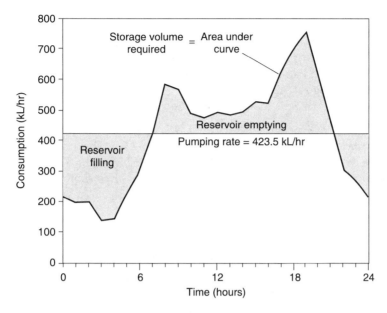

Figure 9.12. Hourly consumptions and pumping rate

Table 9.4. Calculation using sizing method 1

Hour	Consumption (kL/hr)	Inflow-consumption (kL/hr)	Hour	Consumption (kL/hr)	Inflow-consumption (kL/hr)
1	196.7	226.833	13	483.7	-60.167
2	196.7	226.833	14	492.8	-69.267
3	136.3	287.233	15	529.1	-105.567
4	144	279.533	16	522.3	-98.767
5	227.1	196.433	17	622.3	-198.767
6	302	121.533	18	697.2	-273.667
7	415.6	7.933	19	756.2	-332.667
8	583.6	-160.067	20	606.4	-182.867
9	567.8	-144.267	21	454.2	-30.667
10	486	-62.467	22	302	121.533
11	472.4	-48.867	23	265.7	157.833
12	492.8	-69.267	24	211.9	211.633

Sizing Method 2

- Calculate the absolute values of the hourly differences between the pumping rate (inflow) and hourly consumptions
- Add all these values together (which is total area of the curve around the pumping rate line)
- The required size is the sum of all the values divided by '2'

For the data provided in Table 9.3, absolute values of 'inflow-consumptions' are calculated as shown in Table 9.5. The sum of all these absolute values of 'inflow-consumption' is 3674.67 kL. Therefore, the required tank size is 3674.67/2 = 1837.34 kL, which is same as what was found through sizing method 1.

Sizing Method 3

- Calculate the differences between the pumping rate and hourly consumptions, which are the same as columns 3 and 6 in Table 9.4
- Calculate the cumulative values of all the differences in new columns
- The required tank size is the 'maximum value – minimum value'

Table 9.5 Calculation using sizing method 2

Hour	Consumption (kL/hr)	Absolute value of 'Inflow-consumption' (kL/hr)	Hour	Consumption (kL/hr)	Absolute value of 'Inflow-consumption' (kL/hr)
1	196.7	226.833	13	483.7	60.167
2	196.7	226.833	14	492.8	69.267
3	136.3	287.233	15	529.1	105.567
4	144	279.533	16	522.3	98.767
5	227.1	196.433	17	622.3	198.767
6	302	121.533	18	697.2	273.667
7	415.6	7.933	19	756.2	332.667
8	583.6	160.067	20	606.4	182.867
9	567.8	144.267	21	454.2	30.667
10	486	62.467	22	302	121.533
11	472.4	48.867	23	265.7	157.833
12	492.8	69.267	24	211.9	211.633

For the data provided in Table 9.3, Table 9.6 show the calculations using sizing method 3. In Table 9.6, columns 2 and 4 are same as columns 3 and 6 in Table 9.4 respectively. In Table 9.6 columns 3 and 6 show the cumulative difference values.

From columns 3 and 6 of Table 9.6, the maximum value is 1346.33 kL and the minimum value is -491.0 kL. Therefore, the required tank size = 1346.33 – (–491.0) = 1837.33 kL.

9.6 Pipe System Analysis and Design

Water supply pipe system is analysed using the energy equation and applying it within to sections (upstream and downstream). As the loss of energy for such pipes is significant, energy losses are usually considered for such analysis. Figure 9.13 shows a typical water supply pipe system having two pipes. In the figure, HGL stands for 'Hydraulic Grade Line', which accounts for the elevation head and pressure head, whereas the 'Energy Line' accounts for all the heads (elevation head, pressure head and velocity head). Among the losses, the main component is friction loss, which is due to the friction between fluid flow and pipe. In addition to friction losses, there are some more minor losses, which are entry loss, exit loss and loss due to the change of pipe diameter.

Table 9.6. Calculation using sizing method 3

Hour	Inflow-consumption (kL/hr)	Cumulative difference	Hour	Inflow-consumption (kL/hr)	Cumulative difference
1	226.833	226.833	13	-60.167	801.229
2	226.833	453.666	14	-69.267	731.962
3	287.233	740.899	15	-105.567	626.395
4	279.533	1020.432	16	-98.767	527.628
5	196.433	1216.865	17	-198.767	328.861
6	121.533	1338.398	18	-273.667	55.194
7	7.933	1346.331	19	-332.667	-277.473
8	-160.067	1186.264	20	-182.867	-460.34
9	-144.267	1041.997	21	-30.667	-491.007
10	-62.467	979.53	22	121.533	-369.474
11	-48.867	930.663	23	157.833	-211.641
12	-69.267	861.396	24	211.633	-0.008

Pipe friction loss is calculated using the widely-used Darcy-Weisbach equation (some other equations, such as Hazen-Williams and Manning's equations, are also used, however, as these are not so common, they are not discussed here). The Darcy-Weisbach equation is presented as follows:

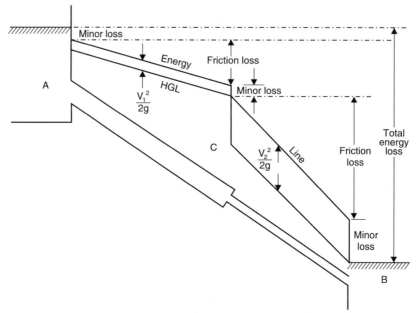

Figure 9.13. Typical energy losses in a pipe supply system

$$h_f = f \frac{L}{D} \frac{V^2}{2g} \tag{9.5}$$

where h_f is the friction loss L is the pipe length, D is the pipe diameter, V is the velocity of fluid within the pipe and f is the pipe friction factor. f depends on Reynold's number of the flow and the pipe roughness coefficient. Reynold's number is defined as:

$$R_e = \frac{VD\rho}{\mu} \tag{9.6}$$

where V and D are defined earlier. ρ is the fluid density and μ is the dynamic viscosity of the fluid. Knowing Reynold's number and pipe roughness, the friction factor (f) can be obtained from the Moody diagram (Figure 9.14), which is widely used in engineering fluid mechanics.

   Exit and entry losses are calculated using exit and entry loss coefficients (K_{ent}, K_{exit}), and multiplying them with the corresponding pipe velocity. Consider the flow from an open tank/pool to another open tank/pool via connecting pipes of different diameters, as shown in Figure 9.15. At equilibrium, the pressure P_1 and the velocity V_1 at point A are zero. Similarly, the pressure P_2 and the velocity V_2 at point B are also zero. From the shown datum, Z_1 is the elevation head of point A and Z_2 is the elevation head of point B. Applying energy equation, $Z_1 = Z_2 + H$, where H is the total energy loss from point A to point B.

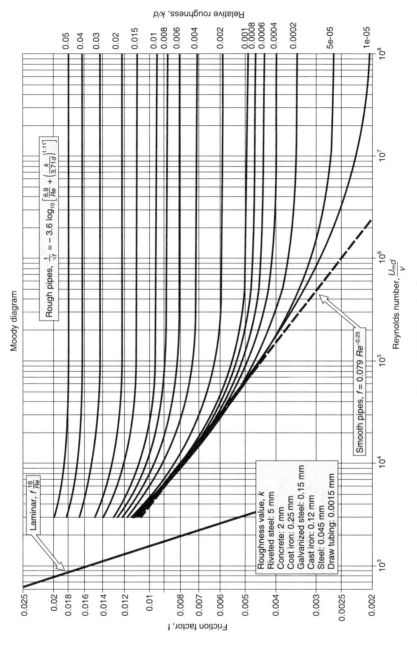

Figure 9.14. Moody diagram

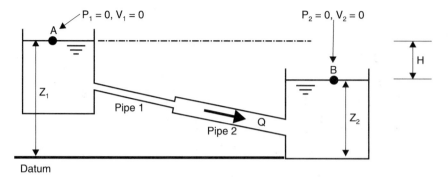

Figure 9.15. Example of losses with static heads

Summing up all the losses, H = entry loss to 'pipe 1' + friction loss in 'pipe 1' + loss due to diameter change from 'pipe 1' to 'pipe 2' + friction loss in 'pipe 2' + exit loss from 'pipe 2' to final chamber. If V_A and V_B are the flow velocities within 'pipe 1' and 'pipe 2', respectively, then the total head loss (H) can be written as:

$$H = K_{ent}\frac{V^2}{2g} + f_1\frac{L_1}{D_1}\frac{V_A^2}{2g} + \frac{(V_A^2 - V_B^2)}{2g} + f_2\frac{L_2}{D_2}\frac{V_B^2}{2g} + K_{exit}\frac{V^2}{2g} \qquad (9.7)$$

In the above equation, if discharge (Q) is known, then V_A and V_B can be calculated as:

$$V_A = \frac{Q}{\pi D_1^2/4}, \; V_B = \frac{Q}{\pi D_2^2/4}$$

For the water supply system, Q is usually known. f values can be calculated with the use of the Moody diagram, applying appropriate values of pipes' roughness. V_A and V_B are calculated from the known Q, as mentioned earlier. K_{ent} and K_{exit} values are usually known from prior experiments. Using the corresponding pipes' diameters and lengths, the total head loss can be calculated using Equation 9.7.

In a different scenario, if H is known while Q is unknown, then Q is assumed and all other parameters are calculated. Then, the calculated H value is compared with the actual H value. The process is repeated until the calculated value is the same as the given value.

9.7 Water Quality

Water can be defined as a tasteless, colourless and odourless liquid composed of Hydrogen and Oxygen. Its chemical formula is H_2O. In addition to direct consumption (which has no alternative), water has several uses in terms of human activities, such bathing and washing, waste disposal (wastes generated from the human body are transported with the aid of water), agricultural uses (i.e. irrigation), industrial uses (i.e. cooling and mixing for many industrial processes) as well as construction and urban development (i.e. mixing of construction material such as concrete). Moreover, water is an essential part of the ecosystem, affecting lives of many other creatures. As such, the demand for water in different sectors has been increasing day by day. It is obvious that the quality of water is as important as quantity, however the required quality depends on the intended use. For example, water used for construction is not required to be of high standard, similarly, water used for industrial cooling does not need to be of drinking water standard. Table 9.7 shows the average proportions of water uses in Australian single houses, including the required quality for each use.

Water can be termed as an almost universal solvent as most of the natural and anthropogenic substances are soluble in it to some extent. Due to it being a universal solvent, water generally contains dissolved substances, many of which can be referred to as impurities. Impurities of water can be classified as: 1) Physical, 2) Chemical, and 3) Biological. Amount and/or concentrations of impurities/contaminants depend on the source of water; surface water often contains suspended matter, whereas groundwater is usually free from suspended matter due to its natural filtration through porous media. Furthermore, surface water is more vulnerable to biological contaminants due to its exposure to the natural environment as well as other animal/aquatic lives. On the other hand, groundwater is usually free from biological contaminants.

Table 9.7. Typical water uses and required standard

Water use	Amount	Potential contact	Required standard
Kitchen	8%	Direct contact	Very high
Bathroom	20%	Contact	High
Laundry	13%	Low contact	Medium
Toilet	15%	Least contact	Low
Outdoor	44%	Least contact	Low

Some potential physical impurities of water are:

- Solids (total, suspended and dissolved)
- Turbidity (reduction in clarity of the water)
- Colour (dissolved organic colloidal material)
- Taste and Odour (decomposed organic matter)
- Temperature

Some potential inorganic chemical impurities of water are:

- pH (measure of acidity or basicity of the water)
- Alkalinity (measure of the capacity to neutralise acids)
- Carbon Dioxide (can be corrosive to pipework)
- Dissolved Oxygen, DO (biological decomposition uses dissolved oxygen)
- Nitrogen (promotes the growth of algae in natural waters)
- Phosphorus (promotes the growth of algae in natural waters and generates from biological decomposition of organic matter)
- Hardness (hard water common in areas with limestone, however generally safe for human consumption)
- Conductivity (measures concentration of dissolved solids)

Some biological impurities/contaminants of water are:

- Bacteria
  - o Simple, colourless single-celled plants
  - o Aerobic bacteria (require free oxygen)
  - o Anaerobic bacteria (Oxidise organic matter in the complete absence of DO)
- Fungi (microscopic non-photosynthetic plants)
- Algae (microscopic photosynthetic plants)
- Protozoa (single-celled aquatic animals)
- Viruses (obligate parasitic particles)
- Pathogens
  - o Disease-producing agents in faeces of infected people
  - o Includes viruses, bacteria, protozoa and helminths (parasitic worms)

9.8 Water Treatment Processes

To treat raw water to the standard of drinking water is a tough, expensive process and requires continuous monitoring of treated water since the lack of proper treatment could result in a potentially catastrophic release of contaminants into the water supply. Some chemical and biological contaminants in particular can cause enormous health hazards. In the current world, approximately 2.2 million deaths occur annually from diarrhoea, among which most are children under 5 years of age. Which means, 1 child dies every 15 seconds, which is equivalent to 20 Jumbo jets crashing every day. Moreover, 10% of the population of developing countries are infected with intestinal worms, worldwide 6 million people are blind from Trachoma and 200 million people are infected with Schistosomiasis. Contaminated water contributes greatly towards these diseases. As such, the treatment of potable water is of utmost importance. Nevertheless, different levels of treatment are required for different intended uses and allowable tolerance levels of contaminants vary. The traditional practice was to treat all the water to drinking water standard. However, with the ever-increasing pressure on potable water supply and scarcity of raw sources, many cities around the world are adopting a dual water supply system or supplying recycled water in addition to a normal potable water supply. If the recycled water is used for outdoor irrigation, car wash and/or toilet flushing only, then it does not need a high level of treatment. If the recycled water is used for laundry, it will require higher level of treatment. In fact, the level of treatment is decided by the potential contact of water with humans. For consumption (drinking or cooking), the level of treatment must ensure elimination of all the harmful elements and bacteria. The treatment processes can be categorised into two major categories; 1) Physical treatment and 2) Chemical treatment. In general, physical treatment involves the change in quality by physical forces/ barriers and chemical treatment involves change in quality by a chemical reaction.

Some common physical treatment methods are:

- Screening
- Aeration
- Mixing
- Flocculation
- Sedimentation
- Combined flocculation & sedimentation
- Filtration
- Membrane processes

Screenings are designed to remove large objects and suspended debris. Coarse screens (20 to 50 mm spacing) are used for bigger particles and small plants and can be cleaned manually. Micro-screens (23 to 65 μm) are used to trap finer particles and are required to be continually washed in order to stop clogging. The most common form of aeration is injecting air from the bottom of the water column or water body. There is yet another simpler form of aeration, which simply conveys water above stepped cascades. Aeration provides multiple benefits; it oxidises dissolved Iron and Manganese, removes Hydrogen Sulphide (i.e. rotten egg gas) and removes volatile oils (some of which cause odour and abnormal taste). Though termed as physical treatment, mixing is mainly used for chemical treatment processes in order to uniformly mix the chemicals used for water treatment. Mechanical mixers are used for this purpose, which is also used to uniformly mix oxygen within the water. Flocculation is a combination of physical and chemical treatments, as for effective flocculation additional chemicals are required. Finer and colloidal particles do not settle on their own and coagulating chemicals help to form flocs, i.e. small floc particles collide with other particles and form larger floc particles. Larger floc particles then can be easily removed by normal settling.

Sedimentation is the most basic and universally used process in physical treatment. As the suspended inorganic particles have a higher density ($\rho \sim 2{,}650$ kg/m^3) in comparison to water ($\rho \sim 1{,}000$ kg/m^3), sediments naturally settle at the bottom of the tank/reservoir, these can be collected at the bottom of the tank and easily removed. The rate of settling depends on the density and the viscosity of the water. It also depends on the size, shape and density of particles. Natural sedimentation of particles through gravity force is often affected by the upward movement of water (due to wind, temperature difference/convection etc.) and non-sphericity of particles. Filtration is another basic and commonly-used process in physical treatment. Filtration uses several layers of media, containing sand or a mix of sand and crushed coal supported on a layer of gravel. Flow travels through layers causing suspended particles and floc attach to sand particles mainly due to the straining action. A common drawback of such filtration system is that pores gradually get clogged over time, causing an increased hydraulic head loss. Typical remedy of such clogging is back-washing of filter media, which removes trapped material. Back-washing is mainly forcing the water back (from opposite direction) through filter media. A common side-effect of such back-washing is the expansion of sand bed material by up to 50%.

Membrane filtration is a contemporary physical treatment process, used for water with high levels of chemical and/or biological contaminations. As membrane filtration uses higher technology with expensive membranes, water should go through other primary forms of pre-treatments (to remove larger particles and pollutants) before it goes through this process. Below are the different types of membrane filtration processes:

- Reverse Osmosis
    - o Pressure driven process
    - o Semipermeable membrane is used for separation
    - o Retains ions and other micro-sized pollutants
- Ultrafiltration
    - o Before real filtration, pressure is used to concentrate colloidal solutions
- Nano-filtration
    - o Low pressure reverse osmosis
    - o Effective in removing hardness, viruses, bacteria and organic colour

In regards to chemical treatments, the following are the main processes:

- Coagulation
- Disinfection

- Water Softening
- Desalination

Coagulation is used where suspended pollutant/material is fine and colloidal. Special coagulants react with water and fine particles and form floc. Then, flocs are converted to large particles, which start settling due to gravity. Alum is the most common coagulant, which forms Aluminum Hydroxide floc.

Disinfection is the most vital water treatment step for potable water supply. Although, if a storage reservoir is used as a part of the urban water supply, 50% of the bacteria die out within 2 days (due to lack of food and/ or oxygen). However, there are some bacteria which can still survive even for more than 2 years. As such, it is necessary to completely annihilate harmful bacteria before it is supplied to the residents for potable uses. Common methods of disinfection are:

- Chlorination
- Ozonation
- UV Radiation, and
- Adsorption

Chlorination is the addition of Chlorine into the water and it has immediate effects on most microscopic lives, including bacteria. It produces Hypochlorous acid and Hypochlorite ion, the combination of these are defined as free available Chlorine. It is an acceptable limit if concentration of free Chlorine in the residual water after 10 minutes of application is 0.2 mg/L. Concentration more than this produces an undesirable taste, however, as Chlorine is cheap, many developing countries tend to use higher amounts of chlorine in potable water in order to be on the safe side in regard to potential pathogens/bacteria. Chloramines are produced when Ammonia is present in the water, which is a slow acting disinfectant and continues disinfecting within the distribution system. As chlorination can produce Trihalomethanes (THM), which can cause cancer, many developed nations are trying to avoid using chlorine. Recently, two new disinfection processes have evolved, which are ozonation and UV radiation, both are powerful and effective for disinfection. They are often used as a secondary disinfectant. For the removal of taste and odour, adsorption using activated carbon is an effective process. Due to its high degree of microporosity, just one gram of activated carbon has a surface area of approximately 500 m^2, which effectively helps to remove taste and odour, which are mainly produced from dissolved Hydrogen Sulphide, living and decaying organic matter (algae), industrial wastes and chlorine.

Water softening is the process of removing hardness; however, this is not an essential element of water treatment as hard water is not harmful for human consumption. Water softening reduces soap consumption and reduces maintenance costs for plumbing. Treatment processes for water softening are: 1) Precipitation and 2) Ion exchange. Precipitation involves adding lime and soda ash to the water, which react with Calcium and Magnesium salts and form insoluble precipitates, which are removed by sedimentation. Ion exchange is basically the exchange of cations with the help of ion-exchanging resin. In this process, Calcium and Magnesium cations are exchanged for Sodium cations as Sodium salts do not cause hardness.

Desalination is the separation of salt from salty water (i.e. sea water). There are several processes through which desalination can be achieved:

- Distillation: Water is boiled and some volatile impurities are released. Water vapour emanates leaving dissolved solids. Later, the water vapour is condensed, which is pure from salts and impurities.
- Freezing: Water temperature is lowered to form ice crystals, then the ice crystals are separated from the brine solution. Ice crystals are warmed to extract pure water.
- Ion-Exchange Demineralisation: Is a similar process to softening, which exchanges ions of minerals with the aim of replacing Sodium cations. However, this process is very expensive for sea water.
- Electrodialysis: This process diffuses selective ions through selective membranes. These selective membranes are selectively permeable to different ions. Following the passing of selective ions, demineralised water is left behind.
- Reverse Osmosis: Through exerting high pressure (10,000 kPa) this method forces high salt content water to pass through a membrane to a chamber of low salt content water. While passing through the membrane, the salt ions are trapped.

9.9 Water Quality Measurement and Calculations

The concentration of impurities/contaminants are generally measured as mass per unit volume of liquid, and the most common measurement is milligram per litre (mg/L), which is equivalent to g/m^3 (gram per metre cube). Very small concentrations are often measured in μg/L, which is equal to 1×10^{-3} mg/L. Another unit also often used is called ppm (Parts per Million), which is defined as:

$$\text{ppm} = \frac{\text{mg/L}}{\text{Specific Gravity of Liquid}}$$

In water resources, a common phenomenon is mixing of liquids of two different qualities, which requires the calculation of an average concentration of pollutant(s) after mixing. The general formula for calculating the average concentration after mixing is:

$$\overline{C} = \frac{V_1 C_1 + V_2 C_2}{V_1 + V_2} \tag{9.8}$$

where V_1 and V_2 are the volumes of the water inflows having pollutant concentrations of C_1 and C_2 respectively.

9.10 Settling of Particles in a Fluid

Particle settling dynamics in fluid/water are important as many of the contaminants are in non-dissolved particulate form, suspended in the water. Through applying proper suspension/settlement dynamics, it is possible to work out how much contaminants can be removed from the water system through a particular sedimentation process. If the particle suspended in the fluid has a lower density than the fluid itself, then the particle will rise/float. If the particle has the same density as the fluid, then the particle will remain stationary and if the particle is more dense than the fluid, then it will settle in the fluid. However, convection and other hydraulic effects may affect this process. In order to perform particle settlement analysis, some assumptions are implied:

- Particles are spherical
  - o In reality particles are not spherical
  - o Assuming spheres match similar spherical behaviour
  - o Only a disk shaped particle will not match with spherical behaviour
- Fluid is homogeneous
  - o No variation in density with depth. This is not true for a deep fluid.
- Fluid is at rest
  - o Convection and other effects are assumed negligible
- Particles are denser than water
  - o Less dense particles will never settle

Considering a particle submerged in a fluid, two forces are acting on it:

Weight force, $W = \rho_s g * \text{Volume} = \rho_s g * \dfrac{\pi D^3}{6}$, and

Buoyant force, $F_B = \rho_w g * \text{Volume} = \rho_w g * \dfrac{\pi D^3}{6}$

where ρ_s is the density of the particle, ρ_w is the density of water, g is the gravitational acceleration and D is the particle diameter.

Taking sum of all the forces in vertical direction and considering upward as positive, $\sum F_{vert} = F_B - W$, which yields,

$$\sum F_{vert} = \rho_w g \frac{\pi D^3}{6} - \rho_s g \frac{\pi D^3}{6} = \frac{\pi D^3}{6} g (\rho_w - \rho_s) \qquad (9.9)$$

If ρ_w is less than ρ_s, then the above term is negative and the particle will start to accelerate downwards. Once the particle starts settling, an additional force (drag force) will act on it as shown in Figure 9.16.

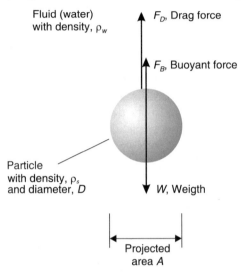

Figure 9.16. Forces acting on a spherical settling particle

The drag force is defined as:

$$F_D = C_D A \rho_w \frac{V_s^2}{2} \qquad (9.10)$$

where C_D is the drag coefficient, V_s is the particle settling velocity and A is the projected area of the particle. Replacing projected area in the Equation 9.10 yields,

$$F_D = C_D \rho_w \frac{V_s^2}{2} \frac{\pi D^2}{4} \qquad (9.11)$$

Now, all the forces acting on the particle can be summed up as (considering upward as positive):

$$\sum F_{vert} = F_B - W - F_D \qquad (9.12)$$

Particles continue to accelerate until equilibrium of the forces occur. Equilibrium will occur only when the sum of all the forces becomes zero, which renders a uniform settling velocity of the particle. At equilibrium, the sum of all the forces can be written as:

$$\sum F_{vert} = \rho_w g \frac{\pi D^3}{6} + C_D \rho_w \frac{V_s^2}{2} \frac{\pi D^2}{4} - \rho_s g \frac{\pi D^3}{6} = 0 \qquad (9.13)$$

The equation can be simplified as:

$$\rho_w g \frac{D}{6} + C_D \rho_w \frac{V_s^2}{8} - \rho_s g \frac{D}{6} = 0$$

$C_D \rho_w \dfrac{V_s^2}{8} + \dfrac{gD}{6}(\rho_w - \rho_s) = 0$, which eventually yields,

$$V_s^2 = \frac{8}{6} \frac{gD}{C_D} \frac{(\rho_s - \rho_w)}{\rho_w} \qquad (9.14)$$

In the above equation, drag coefficient as a function of the Reynolds number, which is shown in Figure 9.17, and Reynold's number is defined as,

$$N_R = \frac{\rho_w D V_s}{\mu_w} \qquad (9.15)$$

It is found that for Reynold's Number <0.5, the nature of flow is laminar and C_D can be simplified as, $C_D = 24/N_R$. Replacing this value into Equation 9.14, George Stokes simplified the settling velocity equation (for laminar flow) as follows:

$$V_s^2 = \frac{8}{6} \frac{gD}{24} \frac{V_s D \rho_w}{\mu} \frac{(\rho_s - \rho_w)}{\rho_w}$$

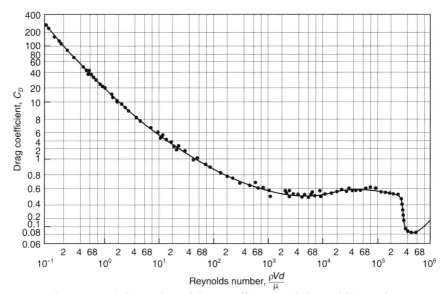

Figure 9.17. Relationship of drag coefficient with Reynold's number

Which can be simplified as:

$$V_s = \frac{1}{18} \frac{gD^2}{\mu} (\rho_s - \rho_w) \qquad (9.16)$$

This is the simplified settling velocity equation, which is valid for $N_R < 0.5$.

For higher Reynold's numbers, the value of C_D has to be extracted from Figure 9.18. The solution process requires trial and error, as explained below:

- Step 1 – Guess settling velocity
- Step 2 – Calculate Reynold's number using Equation 9.15
- Step 3 – Determine drag coefficient from the Figure 9.17
- Step 4 – Solve Equation 9.14 to find settling velocity
- Step 5 – Repeat the same process until assumed settling velocity is equal (or very close) to calculated settling velocity in Step 4

9.11 Sedimentation Basin Sizing

Hydraulic Retention Time (HRT) is the most important parameter in the sizing of a sedimentation basin. HRT is defined as the amount of time water/inflow spends in the treatment process (i.e. in the basin). The mathematical expression of HRT is,

$$HRT = \frac{Vol}{Q} = \frac{A * d}{Q} \qquad (9.17)$$

Where *Vol* is the volume of the basin (L^3), Q is the flow rate (L^3/T), A is the surface area of the basin (L^2) and d is the depth of the basin (L). HRT is expressed with the unit of time (T), usually in either hours or days.

Consider sediment-laden water is entering into a sediment basin of area A with a flow rate of Q. Figure 9.18 shows the plan view (top) and

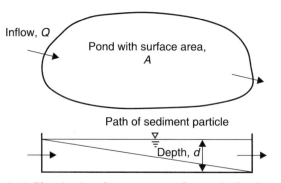

Figure 9.18. Plan (top) and cross-section (bottom) of sediment basin

cross-section (bottom). Depth of the sediment basin is d. The cross-section at the bottom of the figure is showing the sediment travel path as the water travels downstream. To be able to trap the sediment within the tank, the sediment has to settle at the bottom before the inflow water gets out of the tank. If the settling velocity is V_s, then the time required for a particular sediment to settle a distance d (i.e. depth), is equal to d/V_s. As such, for the design purpose to aim to trap a particular sized sediment, HRT should be equals to d/V_s. Using Equation 9.17,

$$HRT = \frac{A*d}{Q} = \frac{d}{V_s}$$

As such, the surface area, $A = \dfrac{Q}{V_s}$ (9.18)

A compromising selection of A, Q and V_s is often crucial. With the ever-increasing urbanisation, authorities want to treat more polluted runoff (Q); to treat more runoff, more surface area is needed. However, the area is limited due to scarcity of land or land cost. To treat a higher Q while having the same area (A), a higher V_s has to be adopted. A higher V_s corresponds to bigger particles/sediments and, as such, the tank/basin will be able to trap only bigger particles corresponding to the adopted V_s.

Worked Example 1

A town in Victoria has a small water supply system that was designed a number of years ago. As part of this system, a 628 kL water supply reservoir was constructed in order to contain the maximum day demand for the town. In 2006, the population of the town was estimated at 950 people. Adopting an exponential population growth model for the town, it has been estimated that the reservoir will no longer be able to supply the maximum daily demand for the town after 2020. Assuming that the average daily consumption in the town is 220 L/person/day, determine the annual growth rate adopted in the analysis.

Solution:

Maximum Day Demand, MD = 628 kL

As per design criteria, MD = PDF*AD

For population less than 2,000, PDF = 2.

So, Average Day Demand, AD = 628/2 = 314 kL

Considering per person daily demand of 220 L, the population for which the AD will be just enough is = 314*1000/220 = 1427.3

Given data: $P_0 = 950$, $P_t = 1427.3$, $t = 2020 - 2006 = 14$ years

Using the Equation 9.2, $P_t = P_0 * e^{Ct}$

$$1427.3 = 950 * e^C * 14$$
$$e^{14C} = 1427.3/950 = 1.502$$
$$14*C = \ln(1.502) = 0.407$$
$$C = 0.029 = 2.9\%$$

So, the annual growth rate is 2.9%.

Worked Example 2

A town in Victoria has a small water supply system that was designed a number of years ago. As part of this system, a 600 kL water supply reservoir was constructed in order to contain the maximum day demand for the town. In 2006, the population of the town was estimated at 950 people. Adopting Logistic Growth model for the town, it has been estimated that the reservoir will no longer be able to supply the maximum daily demand for the town after 2020. Assume the average daily consumption in the town is 220 L/person/day and the population carrying capacity of the town is 10,000. Determine the annual growth rate adopted in the analysis.

Solution:

Maximum Day Demand, MD = 600 kL

As per design criteria, MD = PDF*AD

For population less than 2,000, PDF = 2.

So, Average Day Demand, AD = 600/2 = 300 kL

Considering per person daily demand of 220 L, the population for which the AD will be just enough is = 300*1000/220 = 1363.6

Given data: $P_0 = 950$, $P_t = 1363.6$, $K = 10000$, $t = 2020 - 2006 = 14$ years

Using the Equation 9.4, $P_t = \dfrac{K}{1 + \left[\left(\dfrac{K - P_0}{P_0} \right) e^{-C_{max}t} \right]}$

$$1363.6 = \dfrac{10000}{1 + \left[\left(\dfrac{10000 - 950}{950} \right) e^{-C_{max}*14} \right]}$$

$$\left(\dfrac{10000 - 950}{950} \right) e^{-C_{max}*14} = \dfrac{10000}{1363.6} - 1 = 6.334$$

$$e^{-C_{max}*14} = 0.665$$

$$e^{C_{max}*14} = 1.504$$

$$14*C_{max} = \ln(1.504) = 0.408$$

$$C_{max} = 0.408/14 = 0.0291 = 2.91\%$$

So, annual growth rate is 2.91%

Worked Example 3

A town in Victoria has a small water supply system and as part of the water supply system an overhead tank/reservoir was constructed in order to contain the maximum day demand for the town. The current population of the town is 1000 and the average daily consumption in the town is 220 L/person/day. Adopting Logistic Growth model and an annual population growth rate of 3% for the town, calculate after how many years the reservoir will be no longer able to supply the total demand of the town. The town authority has adopted a maximum population carrying capacity of 10,000, which is to be used for the calculation of future population using logistic growth model.

Solution:

Maximum Day Demand, MD = 600 kL

As per design criteria, MD = PDF*AD

For population less than 2,000, PDF = 2.

So, Average Day Demand, AD = 600/2 = 300 kL

Considering per person daily demand of 220 L, the population for which the AD will be just enough is = 300*1000/220 = 1363.6

Given data: $P_0 = 1000$, $P_t = 1363.6$, $K = 10000$, $C_{max} = 3\% = 0.03$, t is unknown.

Using the Equation 9.4, $P_t = \dfrac{K}{1 + \left[\left(\dfrac{K - P_0}{P_0}\right)e^{-C_{max}t}\right]}$

$$1363.6 = \dfrac{10000}{1 + \left[\left(\dfrac{10000 - 1000}{1000}\right)e^{-0.03*t}\right]}$$

$$1 + [9 * e^{-0.03*t}] = \dfrac{10000}{1363.6} = 7.334$$

$$9*e^{-0.03*t} = 7.334 - 1 = 6.334$$

$$e^{0.03*t} = 9/6.334 = 1.421$$

$$0.03*t = \ln(1.421) = 0.3514$$

$$t = 0.3514/0.03 = 11.71 \text{ years}$$

Worked Example 4

The runoff from a catchment has been estimated as 6.54 m³/s. Suspended solids carried in this runoff are captured in a sedimentation basin, which has a volume of 1,345 m³. The sediment basin was designed to capture sediments having a particle diameter of 75 μm or more. Show that it is reasonable to assume that Stoke's Law is valid in this analysis. Calculate Hydraulic Retention Time (HRT) of the flow and minimum surface area (A) required for this basin. You can assume that the water has a density of 998.0 kg/m³ and a viscosity of 1.002×10^{-3} Pa.s and the sediment has a density of 2,650 kg/m³.

Solution:

Given, $Q = 6.54$ m³/s, $D = 75$ μm $= 75\times10^{-6}$ m, $\mu = 1.002\times10^{-3}$ Pa.s, $\rho_s = 2650$ kg/m³, $\rho_w = 998$ kg/m³.

Assuming Stoke's law, using Equation 9.16, $V_S = \dfrac{1}{18}\dfrac{gD^2}{\mu}(\rho_s - \rho_w)$

$$V_S = \frac{1}{18}\frac{9.81*(0.000075)^2}{1.002\times10^{-3}}(2650 - 998) = 0.00505 \text{ m/s}$$

$$= 5.05 \text{ mm/s}$$

To verify the validation of Stoke's law:

$$N_R = \frac{\rho_w D V_s}{w} = (998*0.000075*0.00505)/1.002\times10^{-3} = 0.378$$

As the Reynold's number is less than 0.5, assumption of Stoke's law is valid.

$$HRT = \text{Volume}/Q = 1345/6.54 = 205.66 \text{ S}$$

Minimum Area required, $A = Q/V_s = 6.54/0.00505 = 1295$ m²

Worked Example 5

In a pond, suspended solids (having average sediments particle diameter of 75 μm) are settling with a fall velocity of 0.005 m/s. Consider that Stoke's Law is valid in this analysis. Calculate the drag co-efficient (C_D) on the sediment particles. You can assume that the water has a density of 998.0 kg/m³ and the sediment has a density of 2,650 kg/m³. Then, calculate the viscosity of the pond water.

Solution:

Given, $V_s = 0.005$ m/s, $D = 75$ μm $= 75\times10^{-6}$ m, $\rho_s = 2650$ kg/m³, $\rho_w = 998$ kg/m³.

Using Equation 9.14,

$$V_s^2 = \frac{8}{6} \frac{gD}{C_D} \frac{(\rho_s - \rho_w)}{\rho_w}$$

$$C_D = \frac{8}{6} \frac{gD}{V_s^2} \frac{(\rho_s - \rho_w)}{\rho_w} = 64.95$$

Now, using Equation 9.16, $V_S = \frac{1}{18} \frac{gD^2}{\mu} (\rho_S - \rho_w)$

$$\mu = \frac{1}{18} \frac{gD^2}{V_S} (\rho_S - \rho_w) = 0.00101 \text{ Pa.s}$$

Worked Example 6

You are engaged in a sediment settlement analysis. It is found that the Reynold's Number for this analysis is higher than 0.50, for which Stoke's law is not valid. The settlement velocity has to be calculated through the trial and error method. Your analysis is for sediments having a particle diameter of 160 μm. You can assume that the water has a density of 1000.0 kg/m^3, a viscosity of 1.000×10^{-3} Pa.s and the sediment has a density of 2,600 kg/m^3. As an initial guess, use a sediment settling velocity of 0.02 m/s. You need to use Figure 9.17. Calculate the sediment settling velocity after the first trial and comment on your initial guess. What would be your second guess?

Solution:

Given, $V_s = 0.02$ m/s, $D = 160$ μm $= 160 \times 10^{-6}$ m, $\rho_s = 2600$ kg/m^3, $\rho_w = 1000$ kg/m^3, $\mu = 1.0 \times 10^{-3}$ Pa.s.

Using Equation 9.15, $N_R = \frac{\rho_w D V_s}{\mu_w}$

So, $N_R = 1000 * 160 * 10^{-6} * 0.02/1.0 \times 10^{-3} = 3.2$

From the Figure 9.17, $C_D = 9.0$

Now, using Equation 9.14,

$$V_s^2 = \frac{8}{6} \frac{gD}{C_D} \frac{(\rho_s - \rho_w)}{\rho_w} = \frac{8}{6} \frac{9.81 * 160 * 10^{-6}}{9} \frac{(2600 - 1000)}{1000} = 0.000372$$

So, $V_s = 0.019$ m/s

As the initial guess is very close to the calculated settling velocity, the next trial can be 0.0195 m/s.

Worked Example 7

The runoff from a catchment has been estimated as 5.0 m^3/s. Suspended solids carried in this runoff are captured in a sedimentation basin, which has a volume of 1,200 m^3 and surface area 1,000 m^2. Considering that Stoke's Law is valid for this analysis, calculate the minimum particle size the sediment basin will be able to capture. What modification is required in order to capture particles having a size smaller than the one found in this calculation. Assume that the water has a density of 998.0 kg/m^3 and a viscosity of 1.002×10^{-3} Pa.s and the sediment has a density of 2,650 kg/m^3.

Solution:

Given, Q =5.0 m^3/s, A = 1000 m^2, μ = 1.002×10^{-3} Pa.s, ρ_s = 2650 kg/m^3, ρ_w = 998 kg/m^3.

Settling velocity, V_s = Q/A = 5/1000 = 0.005 m/s

Using Equation 9.16, $V_S = \dfrac{1}{18}\dfrac{gD^2}{\mu}(\rho_S - \rho_w)$

$V_S = \dfrac{1}{18}\dfrac{9.81 * D^2}{1.002 \times 10^{-3}}(2650 - 998) = 0.005$ m/s

So, D^2 = 0.000000005 m^2

D = 0.0000707 m = 0.0707 mm

So, minimum size which can be captured is 0.0707 mm.

To be able to capture smaller particles one must increase the surface area.

Worked Example 8

4.50 T of water with a concentration of 150 mg/L of suspended solids (SS) is added to a 225 kL reservoir containing water which has a concentration of 20 mg/L of SS. Determine the concentration of SS in the reservoir.

Solution:

Given, V_1 = 4.5 T = 4.5*1000 kg = 4.5 m^3 = 4500 L, C_1 = 150 mg/L, V_2 = 225 kL = 225000 L, C_2 = 20 mg/L.

Using Equation 9.8, final average concentration,

$$\bar{C} = \dfrac{V_1 C_1 + V_2 C_2}{V_1 + V_2}$$

$$= \dfrac{4500 * 150 + 225000 * 20}{4500 + 225000} = 22.55 \text{ mg/l}$$

Worked Example 9

A wastewater treatment plant discharges effluent into a river at a rate of 1.40 ML/day. The effluent has a concentration of 30 mg/L of SS. Upstream of the wastewater treatment plant, the flow and water quality in the river are measured as 5.00 m^3/s and 10 mg/L of SS. Determine the concentration of SS downstream of the treatment plant outlet.

Solution:

Considering a time step of 1 hour (i.e. volume of flows in one hour),

Given, $V_1 = 4.5$ T $= 4.5*1000$ kg $= 4.5$ m^3 $=4500$ L, $C_1 = 150$ mg/L, $V_2 = 225$ kL $= 225000$ L, $C_2 = 20$ mg/L.

Using Equation 9.8, final average concentration,

$$\overline{C} = \frac{V_1 C_1 + V_2 C_2}{V_1 + V_2}$$

$$= \frac{4500 * 150 + 225000 * 20}{4500 + 225000} = 22.55 \text{ mg/l}$$

References

ABS (2004). Water Account Australia, 2000-01, Australian Bureau of Statistics, Cat. No. 4610.0. http://www.abs.gov.au/AUSSTATS/abs@.nsf all primary main features/8C8DD6F104A627DDCA257233001CE4A8? open document

ABS (2017). Water Account Australia, 2015-16, Australian Bureau of Statistics, Cat. No. 4610.0. http://www.abs.gov.au/AUSSTATS/abs@.nsf Lookup/4610.0 Main+Features 12015-16

Wastewater Systems

10.1 Introduction

In an urban setup, a significant quantity of used water is disposed of from our showers, toilets, kitchens, laundry and other amenities. These used waters are called wastewater and are transported along with different types of wastes. Some wastewater is also generated from our garden irrigation, however, traditionally this wastewater is transported to our stormwater drainage system (if the stormwater drainage system is separate from the wastewater drainage system). As such, wastewater contains a higher level of contaminants, including human excreta, and, therefore, requires a higher level of cautions in regards to its collection, transport, treatment and disposal. Because of this sensitivity, wastewater collection pipes usually run deeper into the ground so that any accidental leakage/spillage does not easily reach the ground surface or come into contact with humans. In addition to potential hazards from contaminants, it also causes odour and nuisance to the residents. As such, this is an important component of urban water resources which needs to be dealt with very carefully as its failure may be catastrophic. Due to its higher levels of contaminations, it must not be disposed of in the environment without proper treatments.

Sewage is an alternative term for such wastewater (in some countries stormwater is also termed as wastewater). The system which carries the sewage is called the sewerage system. The sewerage system collects wastewater from households and transports it to the treatment facility. Usually, treatment plants are strategically located in a lower level so that wastewater can be transported by gravity, without the use of any pump or energy. In some cases, if topography does not allow gravity transport, pumped rising mains are used. Eventually, the collected wastewater has to be treated in order to remove harmful contaminants, before it is discharged to the environment (usually in a large water body or ocean). In an urban sewerage system, apart from residential wastewater, wastewater from industrial and commercial establishments can also be disposed of. However, these wastewaters must be only from kitchen/ toilet/shower of those industrial/commercial blocks. Many industrial processes produce several highly contaminated by-products, which require specialised treatment and disposal, as wastewater produced from industrial processes must not be disposed to the urban sewerage system. For such highly contaminated industrial wastes, there are specialised companies who collect, treat and dispose those following special recommended procedures and such waste treatment, disposal is strictly monitored by individual country's environmental regulatory authority (e.g. Environmental Protection Authority in USA and Australia).

10.2 Wastewater Collection System

There are two major types of collection systems; i) Combined system and ii) Separate system. The combined system carries all the stormwater (from roof, open space, road) and wastewater (from households, including toilet, kitchen, shower, etc.) through one pipe system. As such, in order to deal with higher volumes of wastewater, the sizes of the pipes have to be quite large in comparison to the ones in the separate system. In regard to maintenance, the combined system is better, as only one system has to be dealt with and single authority is involved in it. For the combined system, the treatment plant has to be designed to deal with higher volumes of wastewater, however the advantage is that the contaminated waste gets diluted during rainy days. Another disadvantage of the combined system is that if any spillage occurs during a heavy rain event, highly contaminated wastewater will come out and will be a hazard for human health and the environment. An advantage of the separate system is that treatment plants do not need to treat as highly contaminated waste, as the separate stormwater does not need higher/any treatment before it can be discharged into the environment (in some cases the stormwater goes through very basic level of treatments). A disadvantage of the separate system is that two separate systems have to be constructed, monitored and maintained, which can become troublesome, especially when two separate authorities are managing these two separate systems (two systems are managed by two different authorities in the major cities in Australia). On top of that, in many cases, the two pipes are close or crossing very close and, as a result, maintenance/rectification of one requires approval from the other authority. Figures 10.1 and 10.2 show typical schematic diagrams of a combined system and a separate system respectively.

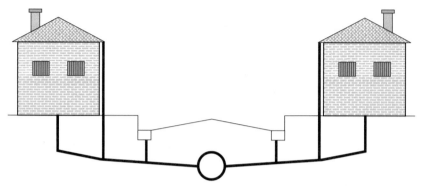

Figure 10.1. Typical schematic of a combined wastewater collection

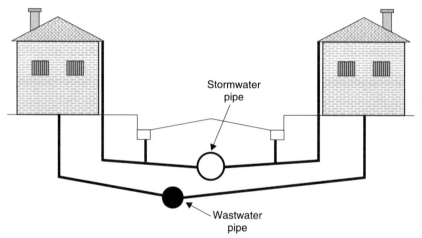

Figure 10.2. Typical schematic of a separate wastewater collections

10.3 Quantification of Wastewater

The volume of wastewater is an important aspect of system design. The components affecting the total volume are: Households, commercial, institutions and industrial. In addition to these common components, another unwanted component also gets in to sewerage system, which is infiltration of stormwater through leaks/joints/cracks and/or illegal connections from the house roof/surface. Sometimes infiltration also occurs when the groundwater table is higher (or the same level) than the pipe level.

Land-use significantly affects the quantity; existing land-use data is obtained from recorded data and future land-use estimates are based on proposed uses. For the purpose of estimation, a unit called 'Equivalent Population or EP' is often used per unit area for different land uses prescribed by the relevant authority. Typical EP loading rates are 180~ 200 L/EP/day. Table 10.1 shows some standard EP values for different land-uses used in Australia.

Table 10.1. Recommended EP values for different land-uses (WSAA, 2002)

Land use type	EP
Single Lot Residential (500 m^2)	50/ha
3 Storey Flats	210/ha
High Density Commercial	500–800/ha floor area
Educational Institution	0.2/student
Hospitals	3.4/available bed
Golf Course	10/ha
Restaurant	500/ha floor area
Hotels/Motels	2,000/ha floor area
Brewery	4,000/ha factory area

For a system design, the design flow is the sum of three components,

$$\text{Design Flow} = \text{PDWF} + \text{IIF} + \text{GWI}$$

where 'PDWF' is the peak dry weather flow. 'IIF' is the peak inflow from infiltration, which is dependent on rainfall, and 'GWI' is the groundwater infiltration, which is not dependent on rainfall, but is dependent on the groundwater level.

Peak Dry Weather Flow (PDWF) is calculated using the following relationship,

$$\text{PDWF} = \text{d} \times \text{ADWF} \tag{10.1}$$

where 'ADWF' is the average dry weather flow, which is usually adopted as 180 L/day/EP (equivalent to 0.0021 L/s/EP). 'd' is the dry weather peaking factor, extracted from Figure 10.3 based on grossed developed area.

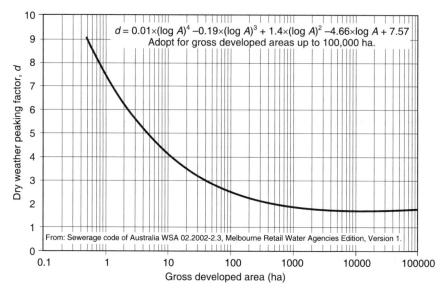

Figure 10.3. Relationship of dry weather peaking factor with gross developed area (*Source*: WSAA, 2002)

10.4 Quality of Wastewater

Wastewater includes contaminants produced by different uses, such as domestic, commercial and industrial. Contaminants found include: i) Physical, ii) Chemical and iii) Biological.

Physical Contaminants

- Solids (Suspended, Volatile, Dissolved)
- Colour
- Odour
- Temperature

Chemical Contaminants

- Chemical Oxygen Demand (COD)
- Total Organic Carbon
- Nitrogen (Ammonia, Nitrates, Nitrites etc)
- Phosphorus (Organic, Inorganic)
- Sulphate
- Heavy Metals
- Various Gases

Biological Contaminants

- Bacteria
  - Range of different types

- o Some are pathogens
- o Some are important in treatment processes
- o Coliforms (usually used as test for harmful bacteria in effluent
- Protozoa
- Helminths
- Viruses

10.5 Wastewater Treatment

Depending on contaminants present in the sewage or aimed to selected contaminants of concern (i.e. suspended solids, organic matter, pathogens and/or nutrients), a range of treatment processes are adopted. Those can be classified as physical, chemical and biological treatments. Again, based on different levels of treatment, the processes are classified as primary, secondary and advanced/tertiary. Eventually, from wastewater treatment, two major effluents come out; one is solid matter from solids and organic matters, which are thickened, dried and disposed of in landfills or incinerated. The other component is liquid (with dissolved solids), which needs to go through proper disinfection processes before it is disposed of in rivers or oceans.

Primary Treatment

This includes physical treatments of wastewater, such as screens to remove large suspended matter, grit removal, shredders and sedimentation. Screens are used to remove suspended matter in wastewater, such as wood, cartons, rags and paper. It is usually used in the upstream portion of the treatment system in order to prevent clogging of downstream machinery. These are usually bar racks, having automatic cleaning mechanisms. The trapped materials are buried, landfilled or incinerated depending on the available facility. Grit removal mainly involves the removal of relatively small particles having specific gravity more than 1.6 (i.e. sand, gravel, eggshells, bones, cigarette butts, etc.). This prevents damage to machinery, as well as accumulation in sludge. There are several different types of systems for grit removals, such as 'Spiral Flow Aerated Chambers' and 'Forced Vortex Chambers'. Trapped grits are usually disposed of in landfills. Primary settling is mainly used to remove larger suspended material, usually high in organic content, having a lower specific gravity (1.2 or less). Due to it having a lower specific gravity, the settling velocity also low (<0.5 mm/s). This tank facilitates sludge to fall to the bottom of the tank, which is usually a sloping floor and allows removal of sludge towards a lowest centric point, when the deposits are flushed out (i.e. removed). Such tanks need uniform velocity distribution at the inlet, which sometimes is attained using baffles. Figures 10.4 and 10.5 show a typical schematic cross-section and plan photo of such primary settling

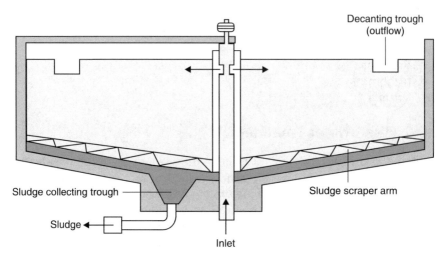

Decanting trough
(outflow)

Sludge collecting trough

Sludge scraper arm

Sludge

Inlet

Figure 10.4. Schematic cross-section diagram of primary settling tank
(*Source*: https://theconstructor.org/environmental-engg/)

Figure 10.5. Plan view of primary settling tank
(*Source*: https://theconstructor.org/environmental-engg/)

tank. Tanks can be of different types; circular, rectangular (having slope to one side).

Secondary Treatment

This includes aerobic biological treatment and final settlement. Biological treatment is done through an activated sludge process, or trickling filters. Under biological filtration, wastewater is put in contact with microbes attached to some media (stone or synthetic material). Microbes start growing since they have enough food (i.e. organic matter in the wastewater) and oxygen. Once the layer of microbes gets bigger and bigger, it is easier to separate from the system.

Trickling filters use the same technique, where stones are kept for such mechanism and wastewater is spread over the stack of stones, where microbial growth occurs on the surfaces of stones. The surface gets thicker and thicker with the presence of enough organic matter and oxygen. However, at one stage, when it becomes very thick, the inner layer gets out of oxygen contact and eventually dies due to lack of oxygen. The whole thick layer then comes out of the stone surface, which is then separated from wastewater. Figure 10.6 shows a formation of biofilm and exchange of nutrients/ions on a solid media while wastewater is passing by.

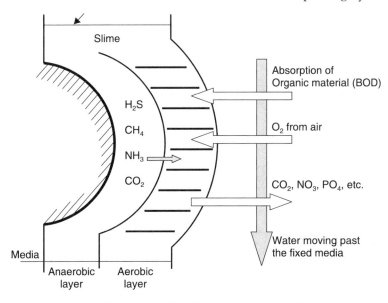

Figure 10.6. Formation of biofilm and exchange of nutrients

During secondary treatment, activated sludge treatment is a commonly-used process, where bacteria and protozoa are mixed with wastewater in liquid suspension. The mixture is aerated in order to promote micro-organism growth. Heavily grown micro-organism mixed with liquor is transferred to the tank which is located after the primary settling tank. Higher number of micro-organisms consume higher organic matters from the wastewater, on the other hand these grown micro-organisms form biological floc, which easily settles in the secondary settlement tank.

Biological Decomposition of Organic Matter

Organic matters in the nature/water naturally decompose by themselves, as long as there are bacteria and enough oxygen. Such decomposition of organic matter can be described with the following chemical equation:

$$\text{Organic (CHNSP)} + O_2 \xrightarrow{\text{Bacteria}} CO_2 + H_2O + NO_3 + SO_4 + PO_4$$

Where, with the aid of oxygen, some bacteria convert organic matter in to CO_2, H_2O, NO_3, SO_4 and PO_4.

The biological decomposition process can be described in Figure 10.7, where the initial phase is the lag phase when the bacteria is not matured enough to divide/grow. With the aid of enough food (organic matter) and environment, bacteria start growing exponentially in the exponential phase. Then, in the stationary phase, the exponential growth ceases due to any growth limiting factor (i.e. depletion of an essential nutrient, formation of an inhibitory organic acid or food). In this stage, the growth rate and the death rate are equal. Finally, during the death phase, bacteria start depleting due to the lack of food and consumption of their own tissue.

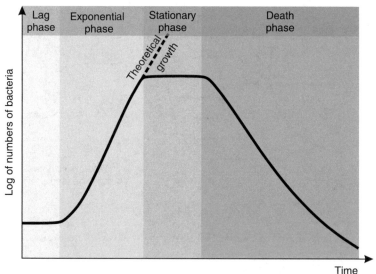

Figure 10.7. Phases of bacterial growth and death
Source: http://library.open.oregonstate.edu/microbiology/

The overall decomposition process can be summarised as three main activities:
- Portion of organic matter is oxidised, which serves energies for cell maintenance and production of new cells
- Conversion of organic matter to new cell tissue, part of the energy from oxidation used here

- Cells consume their own tissue: When all the organic matter is consumed, cells start consuming their own tissue (equivalent to dieting for humans)

In this process, oxygen is required for the oxidation of waste by bacteria. For the organic matter drowned deeper into water, this oxygen comes from the dissolved oxygen in the water, which at times causes depletion of dissolved oxygen in the water, which is harmful for many aquatic lives. This reduction of dissolved oxygen in the water is a very important measure of the amount of organic matter present in the water/wastewater. The measuring unit is named as "Biochemical Oxygen Demand (BOD)", which is the amount of oxygen required for the oxidation of waste (organic matter) by bacteria. This is a very common water quality measuring parameter all over the world. BOD of a water/wastewater sample is the amount of oxygen consumed by the water sample in a total five days at 20° C. Five days being taken as the testing method was developed in England, where five days is generally the longest flow time in their rivers. However, in reality, after five days, the decomposition rate becomes very slow or in other words almost 90% BOD is decomposed. One drawback of this testing method is that, if nitrification is also occurring in the same water/wastewater sample, it also consumes oxygen and, in such case, the real BOD of the sample would be less than what will be revealed by the standard test mentioned above.

BOD Kinetics

Figure 10.8 shows the typical degradation of BOD, as well as total BOD removed with time. This is shown with two opposite lines, as these two are related as per the following equation:

$$BOD_t = L_0 - L_t \qquad (10.2)$$

where BOD_t is the BOD consumed/removed at any time t (usually in days), L_0 is the amount of BOD before oxidation occurs and L_t is the amount of BOD remaining at any time t. As the rate of change of BOD remaining (L_t) is proportional to the total time elapsed since decomposition commenced, the decay curve can be expressed as:

$$\frac{dL_t}{dt} = -k_1 L_t \qquad (10.3)$$

where k_1 is the decay constant (per day).
Integrating Equation 10.3 yields,

$$L_t = L_0 e^{-k_1 t} \qquad (10.4)$$

Therefore, amount of BOD removed/consumed,

$$BOD_t = L_0 - L_t = L_0 - L_0 e^{-k_1 t} = L_0 (1 - e^{-k_1 t}) \qquad (10.5)$$

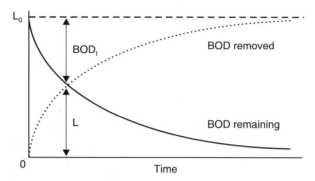

Figure 10.8. Typical BOD decay kinetics

Stabilisation Ponds

Before the final disposal of the liquid effluent after the treatment processes, it passes through a series of stabilisation ponds in order to stabilise organic matter naturally through the combined action of algae and other micro-organisms. Also, in this stage, disinfection may be done (if necessary) by applying the appropriate chemical (i.e. chlorine). The stabilisation process can be wholly aerobic, wholly anaerobic or a mix of aerobic-anaerobic. To achieve a better treatment, a low hydraulic loading rate is preferred, which requires large land areas. In an urban setting, the availability of such large areas is becoming crucial. Figure 10.9 shows the series of stabilisation ponds at Melbourne's Western Treatment Plant.

For the design of stabilisation ponds' area requirement, the following equation is used:

$$A = Q * t / D \tag{10.6}$$

where A is the area required (m^2), Q is the volumetric flow rate (m^3/d),

Figure 10.9. Photo of stabilisation ponds

t is the retention time (day) and D is the pond depth (m). In order to be able to treat a higher amount of *Q*, a bigger area will be needed. Again, to treat to a higher level providing longer retention time, a bigger area will be needed. However, with the increase of depth (*D*), a smaller area will be needed. However, in reality, due to maintenance ease, the depth should not be too great. As such, a prudent decision is required in selecting a combination of these parameters. However, in reality, in an urban setting, the decision is mainly influenced by the lack of a larger area, even though it needs to treat a higher flow volume.

Nutrients Removal

In the urban wastewater, nitrogen and phosphorus are present in different forms. In the past it was not a big issue, however with ever increasing urbanisation and increased uses of different chemicals and detergents, concentrations of nitrogen and phosphorus being discharged with the wastewater are becoming very high. Even after all other treatments mentioned above, such wastewater contains higher levels of nitrogen and phosphorus, and, if released in the surrounding water body will cause eutrophication (enrichment of nutrients which causes algal blooms). The removal of these nutrients require more complex processes like:

- Nitrification (ammonia oxidised to nitrate)
- Denitrification (nitrate converted to nitrogen gas under anoxic conditions)
- Biological phosphorus removal, which is done by growing bacteria that are capable of storing phosphorus in their cells, then removing these phosphorus-accumulated organisms (as sludge)

10.6 Disinfection, Sludge Treatment and Disposal

Final end products of wastewater treatments are, i) liquid portion, which needs to be disinfected before disposal and ii) semi-solid sludge, which needs to be dried/dewatered before being disposed of/landfilled/ stockpiled.

Disinfection basically kills pathogens in the effluent before discharging it into the environment, in order to protect the public as well as aquatic health/lives. A very common test of disinfection success is the tests for Faecal Coliform bacteria. If the concentration this bacteria is below a maximum number (defined by the local regulatory authority, i.e. Environmental Protection Authority), then concentrations of all other bacteria are also expected to be lower and the disinfection is deemed satisfactory. Chlorination (i.e. mixing of chlorine) is the common practice of disinfection. Chlorine is mixed with effluent, and the residual chlorine concentration is measured at the outlet in order to ascertain non-toxic conditions in receiving waters. Another technique is 'ultraviolet radiation', which when absorbed by DNA of micro-organisms, causes prevention of micro-organism reproduction. Usually ultraviolet lamps are used for such purposes.

Sludge is produced in the sedimentation processes of different tank-based treatments, which contains offensive material. It is mainly organic matter from biological treatment. It has a very high water content; only a small part is solid. Different processes of sludge treatment are:

- Digestion
- Dewatering
- Disposal

Sludge digestion is a biological process, which is usually anaerobic. Anaerobic and facultative bacteria convert 40 to 60% of organic matter. The final products produced are Carbon Dioxide and Methane. The remaining organic matter is usually chemically stable, practically odourless, and contains 90 to 95% moisture. Such conversions require warm temperature and methane gas which is used to heat digesters.

Sludge dewatering is the separation/extraction of water from the sludge. The usual practice is to leave the sludge to dry naturally on the sludge drying beds. After such drying, a very fine powder forms from the end product remains. Sometimes, dewatering is also done mechanically and there are two main types of mechanical separation; i) Belt filter process and ii) Centrifuge process. The belt filter process squeezes sludge between porous belts and water comes out of the pores. However, still some moisture remains in the end product. The Centrifuge process uses centrifugal force through rotation in order to separate water from the solid. However, this process is not as common as the belt filter process and is generally used for large facilities.

The final stage is the disposal of dried sludge. The traditional practice was to dump dried sludge into common landfill or in a dedicated safe place. In the dedicated place, the dried sludge is disposed of layer by layer, becoming heaps of sludge. Regulatory authorities do not allow these heaps to go beyond a certain height, due to safety and odour issues. Many urban water/sewerage authorities are struggling to find new places to dump the dried sludge as the existing facilities are already full in their capacities.

10.7 Sustainable Wastewater Treatment and Recycling

With the burden of ever increasing population and consequential generation of wastes, while on the other hand limitation of space and resources, the traditional practices of wastewater treatments are not really sustainable. Authorities and researchers are exploring other sustainable approaches to overcome this issue. One of the sustainable approaches is to use a decentralised sewer system instead of a traditional centralised sewer system. In the traditional centralised system, all the wastewater from the whole (or a major part of the) city is transported and treated in a central location, which requires a very long pipe network, as well as very big treatment facilities, which is a burden for the authority. The decentralised system proposes to treat wastewater in several pocket areas in smaller treatment plants and each plant will collect wastewater from a smaller neighbouring area. This will reduce the requirement of very long pipe systems, as well as very big treatment facilities. Also, these smaller treatment plants will endeavour to reuse their final end products, including dried sludge.

Some smaller cities are treating the final water portion of the wastewater and supplying it to the same locality. However, a high level of treatment is required if it has to be used for potable purposes. Some cities are found to adopt this when affected by severe drought for prolonged periods, even though many residents were not ready to accept such recycled water. The reality is that nature has been gradually recycling all the wastewater through a longer process, called the natural hydrologic cycle, which was discussed in Chapter 1. The water that we are drinking now (or using for other foods) might have been wastewater produced by someone else (or by some city/village) hundreds/thousands of years ago. Considering this issue of residents' acceptability, many new suburbs in Australia adopted a lower level of localised treatment of wastewater and are recycling it for irrigation/outdoor or toilet flushing purposes only.

In regards to dried sludge, it is being used as fertilizer in the potting mixes for application in agricultural land with limited capacity. Some researchers even suggest to use raw sludge for agricultural application. Recently, some researchers conducted engineering and environmental tests on stockpiled dried sludge and reported it to be suitable for soft-engineering structures such as non-load bearing embankment fills. However, for any such reuse, the dried sludge needs to go through proper leaching tests to ascertain that it does not contain harmful pollutants in excess of specified upper limits. These samples in particular are likely to be contaminated with heavy metals, which cannot be removed through traditional treatment processes.

Worked Example 1

A current wastewater treatment pond (volume of existing wastewater 5000 m^3) is holding wastewater having a BOD concentration of 15 mg/l. After 5 days, while this wastewater is getting treated, another wastewater flow entered/discharged into the pond at a rate of 10 l/s for 5 hours with a BOD concentration of 30 g/m^3. Assume that the existing and new wastewaters are well-mixed in the pond and ignore BOD decay within this five hours discharge/mixing period. Determine the BOD concentration in the pond after an additional four days. Consider a BOD decay constant (k_1) of 0.20/day for all the wastewater in the pond.

Solution:

For the wastewater already in the pond,

L_0 = 15 mg/l, k_1 = 0.2/day. Pond volume, V_1 = 5000 m³

So, BOD in the pond after five days,

$$L_5 = L_0 e^{-k_1 * 5} = 15 * e^{-0.2 * 5} = 5.52 \text{ mg/l} = C_1$$

Volume of new wastewater entered = V_2 = 10*5*60*60/1000 = 180 m³,

$$C_2 = 30 \text{ g/m}^3 = 30 \text{ mg/l}$$

So, average concentration,

$$\bar{C} = \frac{5000 * 5.52 + 180 * 30}{5000 + 180} = 6.37 \text{ mg/l}$$

Using Equation 10.4, BOD concentration after an additional 4 days,

$$L_4 = L_0 e^{-k_1 * 4} = 6.37 * e^{-0.2 * 4} = 2.86 \text{ mg/l}$$

Worked Example 2

For a wastewater sample, it has been estimated that BOD after two days is 15 mg/l and BOD after six days is 5 mg/l. Calculate the BOD at the beginning (L_0) and the decay constant (k_1) for the wastewater sample. What would be the BOD concentration after 8 days?

Solution:

L_2 = 15 mg/l, L_6 = 5 mg/l

Applying, Equation 10.4 for these two periods,

$$L_2 = 15 = L_0 e^{-k_1 * 2} \tag{1}$$

$$L_6 = 5 = L_0 e^{-k_1 * 6} \tag{2}$$

Dividing 'Equation 1' by 'Equation 2':

$$\frac{15}{5} = \frac{L_0 e^{-k_1 * 2}}{L_0 e^{-k_1 * 6}} = e^{6k_1 - 2k_1} = e^{4k_1}$$

$\Rightarrow e^{4k_1} = 3$

$\Rightarrow 4k_1 = \ln(3) = 1.099$

$\Rightarrow k_1 = 1.099/4 = 0.275$

Using 'Equation 1',

$$L_2 = 15 = L_0 e^{-k_1 * 2} = L_0 e^{-0.275 * 2}$$

$$L_0 = 15 * e^{0.275 * 2} = 26 \, \text{mg}/l$$

So, BOD after eight days,

$$L_8 = L_0 e^{-k_1 * 8} = 26 * e^{-0.275 * 8} = 2.88 \, \text{mg}/l$$

Worked Example 3

Two wastewater discharge outlets are discharging into a wastewater treatment pond. One outlet is flowing at a rate of 5 l/s with a BOD concentration of 30 mg/l and the other outlet is flowing at a rate of 0.01 m³/s with a BOD concentration of 5 g/m³. Wastewaters from both the outlets are well-mixed in the pond. The pond has a volume of 8000 m³. After staying for the intended retention time in the pond, the wastewater is discharged through a spillway. During the retention time BOD gets decayed. Determine the BOD concentration of the wastewater discharge from the spillway (i.e. after treatment) considering a BOD decay constant (k_1) of 0.20/day.

Solution:

Given, $V_1 = 5$ l/s, $C_1 = 30$ mg/l and $V_2 = 0.04$ m³/s = 10 l/s, $C_2 = 5$ g/m³ = 5 mg/l

Average concentration after mixing,

$$\overline{C} = \frac{V_1 C_1 + V_2 C_2}{V_1 + V_2} = \frac{5 * 30 + 10 * 5}{5 + 10} = 13.33 \, \text{mg}/l$$

Total flowrate = $V_1 + V_2 = 5 + 10 = 15$ l/s

Retention time, t = Volume/flowrate = 8000*1000/15 = 533,333 s = 6.17 days

So, using Equation 10.4, BOD after 6.17 days,

$$L_t = L_0 e^{-k_1 t} = 13.33 * e^{-0.2 * 6.17} = 3.88$$

So, final BOD concentration is 3.88 mg/l.

Reference

WSAA (2002). Sewerage Code of Australia. Melbourne Retail Water Agencies Edition, Version 1.0, WSA 02-2002-2.3. Water Services Association of Australia.

Stormwater Drainage

11.1 Introduction

In a natural/rural catchment, stormwater drainage is not an important issue, as nature has its own way of draining stormwater from catchment surfaces to natural watercourses/creeks through some small natural drains. In most rural cases, separate infrastructure is not needed in order to convey stormwater from the source to a watercourse/waterbody. In addition, a significant amount of stormwater falling on natural ground infiltrates through the ground and does not cause any impact on the earth. However, in an urban setup, stormwater drainage is an important matter. This is due to man-made infrastructures (i.e. road, building, pipe, culvert, pit, etc.) causing significant obstructions to the natural flowpath of stormwater runoff. Moreover, conversions of natural surfaces to impervious surfaces (i.e. roof, road, footpath, etc.) produce significant amounts of stormwater runoff on the surface (as amount of infiltration reduces due to impervious surfaces), which must be conveyed to a downstream location (waterbody) in a safe manner.

For example, Figure 11.1 shows a natural park with contours (land falling from top to bottom), where subdivision and new house blocks are being planned. When the houses are constructed along the road, these houses will experience a huge surface runoff generated from the park, which may cause flooding above the floor for the planned houses. As such, a proper stormwater drainage is required if houses need to be built here. Another example is shown in Figure 11.2, showing a natural surface with contour lines (surface falling from top-right corner towards bottom-left corner), where roads and subdivision of land is being planned.

When houses are built along the roads as shown in Figure 11.3, the rooftops and impervious surfaces will produce a significant additional amount of runoff for the downstream properties. Moreover, these houses need to be safe from runoff flowing from upstream. As such, a

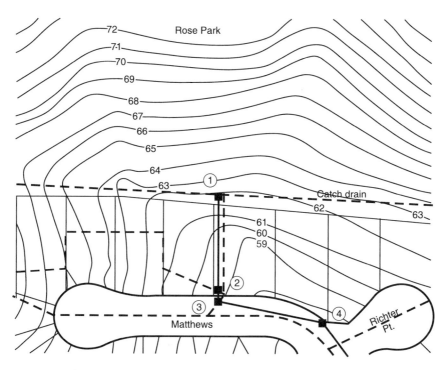

Figure 11.1. Planned subdivision at the end of a natural park

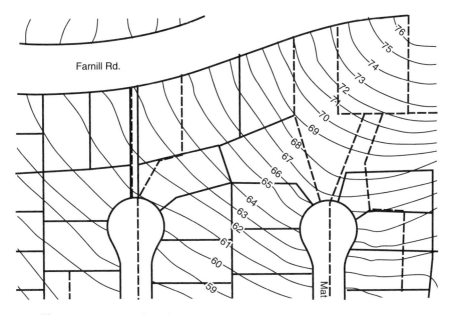

Figure 11.2. Natural surface showing contours and planned subdivision

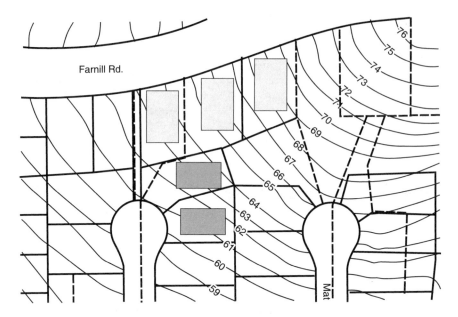

Figure 11.3. Design of house blocks on a natural surface

proper stormwater drainage design is necessary in order to convey the produced runoff downstream in a safe manner. Usually, this is done through underground pipes running along the streets, depicted in Figure 11.1 as thick straight lines connected with square boxes (drainage pits). In general, the analysis of urban stormwater is complicated as it deals with different types of flows, such as impervious surface flow (from road, footpath, driveway), natural surface flow (from natural grassed surface), piped flow (from roof and from road surface to pipe underneath the road) and flow along the street kerb.

11.2 Components of Urban Stormwater Drainage

Below are some basic components of an urban drainage system:
- Lot/house drainage
  - Piped drainage (from roof and from other surfaces to street gutter)
  - Surface drainage (from natural and/or impervious surfaces to street)
- Street drainage
  - Pipe drainage (from street surface, as well as house connections)
  - Kerb (channel) drainage (from house connections and street surface)
- Open channel drainage (from natural surfaces, as well as street drainage pipes)

- Detention basin (may or may not exist depending on locality/land availability)

Figure 11.4 is a photo of a house where the stormwater connection from the roof to the ground through a roof gutter and a downpipe is shown. Several downpipes from a roof (depending on roof size) eventually come underground, join and drain at the street gutter (or sometime directly connected to the underground stormwater pipe along the street, which is not visible). Figure 11.5 shows a photo of a downpipe connection draining to a street gutter. From the Figure 11.5 it is also clear that runoff from pervious and impervious surfaces would drain to the street gutter. Figure 11.6 shows a photo of a pit, through which runoff from the street gutter will enter into underground stormwater pipes. These underground pipes are not visible; Figure 11.7 shows such underground stormwater pipes during laying stage.

Figure 11.4. Roof drainage through gutter and downpipe

Figure 11.5. Connected downpipes draining to street gutter

Figure 11.6. Photo of a pit draining from street to underground pipe

Figure 11.7. Photo of laying of stormwater drainage pipe

Drainage pipes along streets eventually drain to an open channel or pond/lake/reservoir. Figure 11.8 and 11.9 show a stormwater pipe outlet at the beginning of an open channel and towards a pond, respectively. This type of pond (shown in Figure 11.9) acts as a detention pond, which holds the water up to its capacity and releases either continuously with a smaller flow or, when flow is above its capacity, through a weir.

Figure 11.8. Photo of stormwater pipe outlet to an open channel

Figure 11.9. Photo of stormwater pipe outlet to a detention pond
(*Source*: https://www.clemson.edu/extension/water/stormwater-ponds/)

11.3 Design Calculations and Equations

Flow through Downpipes

For the down-pipes, flow to the pipe is either weir flow (when the space above opening is not completely full) or orifice flow (when the space above the opening is completely full, i.e. pressure flow). Smaller diameter pipes at their full capacity are expected to flow abiding the orifice flow equation. In regard to the design of down-pipes, the following website provides an excellent service: http://www.roof-gutter-design.com.au/Downp/applet.php

If roof area, slope and rainfall intensity values are provided through the window shown in the Figure 11.10, the tool provides flow (l/s) generated from the roof for the specified design rainfall. The tool also provides the number of downpipes required to carry this flow with options of different diameter pipes such as 90, 100, 150, 225 and 300 mm, as shown in the Figure 11.11.

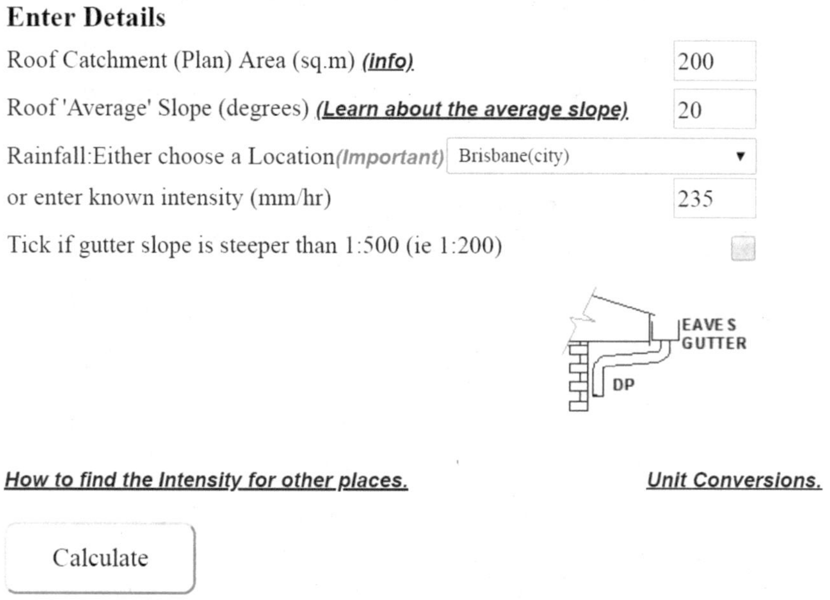

Enter Details

Roof Catchment (Plan) Area (sq.m) *(info)*	200
Roof 'Average' Slope (degrees) *(Learn about the average slope)*	20
Rainfall:Either choose a Location*(Important)*	Brisbane(city) ▾
or enter known intensity (mm/hr)	235
Tick if gutter slope is steeper than 1:500 (ie 1:200)	☑

How to find the Intensity for other places.   *Unit Conversions.*

Calculate

Figure 11.10. Input data menu for downpipe design calculation

Flow along Street Gutter

When roof and surface flows come out into the street gutter, then the gutter flow capacity can be calculated using the Manning's equation discussed in Chapter 7, considering the triangular section between the street kerb and sloped road surface. For gutter flow, below are the critical parameters that must be considered:

- Velocity of flow
- Depth of flow
- Width of flow

The design considerations are that the velocity of flow along the gutter should be safe enough for the pedestrians (especially kids). Depth of flow also should not be higher than a certain depth which becomes unsafe for the pedestrians or vehicles. Also, the width of the flow should not be such that the flow covers the whole street width making the street surface invisible. Figure 11.12 shows a typical street cross-section, where the dashed lines depict safe (lower line) and unsafe (upper line) flow widths.

You will require <u>one</u> of the following DP options :- (dimensions in mm)
(Assuming approximately equal catchment areas)

Flow (L/s) | 15.43

Results:

	Number Req'd	Number Used	*Gutter Area?*	Gutter Width	*Gutter Depth?*
90 Dia:	7.38	8	8268	130	65 ○
100 Dia:	5.82	6	10696	145	75 ○
150 Dia:	2.33	3	19917	200	100 ○
225 Dia:	0.93	1	51652	315	165 ○
300 Dia:	0.49	1	51652	315	165 ○

Figure 11.11. Results of downpipe options for selected roof and rainfall

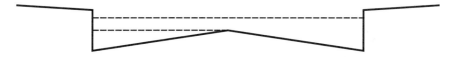

Figure 11.12. Typical street cross-section showing safe and unsafe widths

In regard to flow velocity criteria, a commonly adopted principal is that $v*d \leq 0.4$ m²/s. For a certain cross-sectional property of kerb and gutter (i.e. side slope), considering Manning's n value for concrete, for different combinations of d and v (which satisfies the criteria $v*d \leq 0.4$), flowrate versus longitudinal slope can be drawn (solid line in Figure 11.13). Again, for the same section (same side slope and Manning's n value), for certain allowable depth using Manning's equation, flow versus longitudinal slope curve can be drawn (large dashed line in Figure 11.13). Similarly, for the same section (same side slope and Manning's n value), for a certain allowable flow width, using Manning's equation, flow versus longitudinal slope curve can be drawn (small dashed line in Figure 11.13). (Note: These graphs will be different for different cross-section and gutter material properties). In Figure 11.13, the zone under hatched lines are the safe flowrates which will satisfy all the three criteria.

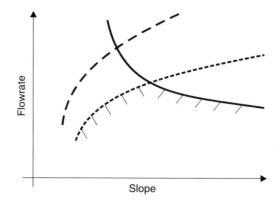

Figure 11.13. Flow capacity chart for different slopes under three criteria

Flow through Street Pits

Along the street, two types of pits are commonly used (in reality there are a few more types used for other different purposes, i.e. junction pit, letterbox pit, etc.); i) Kerb inlet pit and ii) Grated pit. The kerb inlet pit is shown in Figure 11.6, whereas the grated pit (in combination with kerb inlet pit) is shown in Figure 11.14.

The flow equations are different for these different types of pits. For a kerb inlet pit the flow capacity equations are:

$$Q = 1.66*L*d^{1.5} \qquad\qquad (11.1)$$

Figure 11.14. Grated pit in front of a kerb inlet pit

(for ponding up to about 1.4 times the height of the inlet, h, *i.e.* $d = 1.4\,h$) or

$$Q = 0.67*A*[2g(d - h/2)]^{1.5} \qquad (11.2)$$

where, Q is the inlet flowrate (m³/s), d is the average depth of ponding (m), L is the inlet width (m), A is the clear area of the opening (m²), and g is the acceleration due to gravity (m²/s).

For a grated pit, the flow equations are:

$$Q = 1.66*P*d^{1.5}, \text{ up to about 0.12 m of ponding } (d < 0.12) \qquad (11.3)$$

or

$$Q = 0.67*A*(2gd)^{0.5}, \text{ over 0.43 m of ponding } (d > 0.43) \qquad (11.4)$$

where Q is the inlet flowrate (m³/s), d is the average depth of ponding (m), P is the perimeter length of the pit, excluding the section against the kerb (m), A is the clear area (i.e. total area minus area of bars) of the opening (m²), and g is the acceleration due to gravity (m²/s). Note: Here two equations are provided, one for d<0.12 m and the other for d>0.43 m, for depths in between both the equations can be used to calculate separate flowrates, then appropriate interpolation should be done depending on the depth of flow (d).

Pipe capacity

Pipe capacity mainly depends on cross-sectional area, slope and roughness of the pipe material. For simpler analysis, Manning's equation is used to calculate the pipe velocity/capacity. Apart from cross-section and slope, Manning's equation requires a roughness co-efficient (n), which is already established for different materials and discussed in Chapter 7. For a more accurate analysis, the Colebrook-White formula is used to calculate velocity (Colebrook, 1939).

$$V = -2\sqrt{2}\sqrt{gDS}\log\left(\frac{k}{3.7D} + \frac{1.775v}{D\sqrt{2gDS}}\right) \qquad (11.5)$$

where S is the hydraulic gradient (= head loss/length), D is the diameter of the pipe, k is the pipe roughness, v is the kinematic viscosity and g is the acceleration due to gravity.

Usually, the pipe manufacturer provides a flowrate chart for their pipes based on gradient and diameter of the pipe. Australian pipe manufacturer, Rocla (www.rocla.com.au), has provided such charts using both Manning's equation and the Colebrook-White equation (for different k values). As a sample, such a chart using Manning's equation is shown in Appendix B, as provided on the Rocla website.

Open Channel Flow

From the pipes, stormwater usually flows into open channels/creeks. Manning's equation is used for the determination of channel flow capacity, which is discussed in detail in Chapter 7. For a regular channel, the calculation of area is straightforward. However, for irregular channels, knowing the channel bottom levels at equal intervals, the area can be calculated using the 'Trapezoidal rule', as discussed in Chapter 2.

11.4 Hydraulic Grade Line (HGL) Analysis

As discussed in Chapter 9, for water supply pipes, the similar HGL analysis concept is used for stormwater pipe analysis. Figure 11.15 shows a pipe system with three pipes and three open pits. In the figure, TEL stands for 'total energy line'. HGL is the line where the water will reach if it is released from the pipe (i.e. at atmospheric pressure). As discussed in Chapter 6, the total energy/head is expressed by

$$H = \frac{P}{\rho g} + Z + \frac{V^2}{2g} = H_S + \frac{V^2}{2g} \tag{11.6}$$

where H_s is the static head, in other words, HGL is showing the line on static heads along the pipe. In the figure, h_p is the energy loss within the pit (i.e. due to flow obstruction by the pit). Pit energy/head loss is calculated by

$$h_p = K_u * \frac{V^2}{2g} \tag{11.7}$$

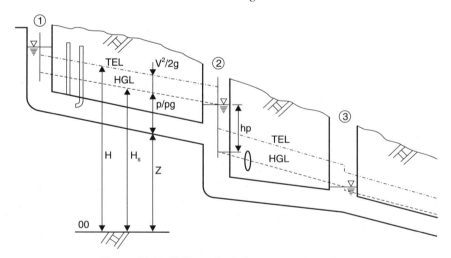

Figure 11.15. HGL analysis for stormwater pipes

where h_p is the head loss (m), K_u is the head loss coefficient and V is the full-pipe velocity (m/s).

Accurate estimation of K_u value requires iterative process. Australian Rainfall and Runoff (Ball et al., 2016) has recommended several initial K_u values for different pipe and grate inflow (Q_g) configurations as shown in Table 11.1. In the Table, Q_o is the outlet/downstream pipe flow and Q_u is the upstream pipe flow.

Table 11.1. Recommended pressure change coefficients for inlet structures

Pit configuration	*Initial K_u*	*Pit sketches*
First pit at the top of a line	4.0	
Well-aligned junction pit with straight through flow, no sidelines, no grate inflow	0.2	
Well-aligned pit with straight through flow, no sidelines, 50% grate inflow	1.4	
Pit with a 90° right angle direction change, no side lines, 50% grate inflow	1.7	
Pit with a straight through flow, one or more sidelines	2.2	
Pit with a right angle direction change from two opposed inflow pipe	2.0	

Another loss is pipe friction loss, which is calculated using the widely-used Darcy-Weisbach equation (as discussed in Chapter 9), which is presented as follows:

$$h_f = f \frac{L}{D} \frac{V^2}{2g} \tag{11.8}$$

where h_f is the head loss (m), f is the friction factor (0.008~0.08), L is the pipe length (m), D is the inner diameter of pipe (m) and V is the velocity of fluid (m/s).

Considering the static heads of 'pit 3' and 'pit 2', the energy equation can be written as (Refer to Figure 11.15):

$$H_2 = H_3 + (h_p)_3 + (h_f)_2 \tag{11.9}$$

where $(h_p)_3$ is the pit energy loss at 'pit 3', $(h_f)_2$ is the pipe friction loss of 'pipe 2' (between 'pit 2' and 'pit 3'), H_3 is the static energy at the end of 'pit 3' and H_2 is the static energy at the beginning of 'pipe 2'. If H_3 is known (or fixed by the authority), then H_2 can be calculated using Equation 11.9, provided that the discharge and friction factors are known. Discharge is required in order to calculate velocity within the pipe and discharge is usually known from catchment analysis. Once H_2 is calculated, the calculation proceeds upstream and H_1 can be calculated using the following equation:

$$H_1 = H_2 + (h_p)_2 + (h_f)_1 \tag{11.10}$$

where $(h_p)_2$ is the pit energy loss at 'pit 2', $(h_f)_1$ is the pipe friction loss of 'pipe 1' (between 'pit 1' and 'pit 2'), H_2 is the static energy at the end of 'pit 2' and H_1 is the static energy at the beginning of 'pipe 1'. Similarly, such calculation can be used for a series of pipes and this is called HGL analysis. In summary, upstream water level (i.e. static head) can be calculated using such analysis, provided downstream water level, discharge, pipe properties (length, diameter, level and friction factor) and pit properties (loss coefficient, level) are known.

11.5 Onsite Detention (OSD) Tank

Rapid urbanisation increased the impervious surface and consequently decreased the subsurface infiltration, which has resulted in an increased risk of urban flooding in many urban areas. The traditional practice was to provide large detention basins downstream of a large catchment. In many cases, with the ever-increasing developments, such large detention basins are not capable of holding runoff coming from upstream up to the design standard. As for example, a detention basin which was designed to hold a maximum flood of 1 in 10 year ARI (Average Recurrence Interval), is no longer capable to hold flood of designed magnitude due to increased impervious surfaces and increased coefficient of runoff. Consequently, such detention basins are experiencing frequent flooding, which are causing enormous damage to the society, environment and economy. Traditionally, the government used to build/provide such detention basins to the residents at no cost to the individual household. But with the rapid increase in urban population and developments, government authorities are no longer capable of providing such large detention basins due to budgetary constraints. In reality, government authorities are shifting this burden to private developers, who are developing a large block of land into many houses and providing a detention basin, large enough to hold a maximum runoff generated from their land from a design rainfall intensity recommended by the local government authority. For such recommendations, the criteria is often that the new developments should not increase the runoff compared to the runoff before development. This facility is named as onsite detention (OSD) basin. Figures 11.16 shows a photo of urban OSD basin, where the front corner of the land surface is integrated with two-sided shallow walls to form an OSD basin. Figure 11.17 shows another photo of an OSD basin, where four sided walls with fences have been integrated with land surface in order to form the OSD volume.

To calculate the OSD volume required, one first needs to calculate runoff generated from the block/house, including impervious and pervious areas. The authority (i.e. council or municipality) recommends a certain design (ARI/AEP) rainfall for a specified duration to be contained within the OSD basin, while permitting a certain discharge to be drained (called permissible site discharge, PSD) to the main drainage system (pipe or street gutter). First, the design rainfall is extracted for the recommended

Figure 11.16. Typical house front corner integrated with OSD

Figure 11.17. Typical OSD integrated with four sided walls and fence

event and specified time period. The design rainfall is then converted to the runoff applying proper runoff coefficient, using the method described in the Chapter 5. For such calculations, the total time period is usually divided into several segments of equal time steps (say five minutes).

The required OSD volume calculation process can be mathematically expressed as follows:

$$SV = \sum_{t=1}^{T}(RV_t - PSDV_t), \text{ if } SV_t < 0, SV_t = 0 \qquad (11.11)$$

where SV is the storage volume required, RV_t and $'PSDV_t$ are the runoff volume and PSD volume, respectively, at any time t. T is the total number of time steps. To maintain the allowable PSD, proper orifice diameter and the OSD height need to be selected. If the required OSD basin height is calculated based on the required volume and basin topography, the required outflow orifice diameter is calculated using the following equation:

$$D^2 = 4 * \text{PSD}/(C_d * \pi * \sqrt{gh}) \qquad (11.12)$$

where D is the required orifice diameter, h is the depth of the OSD chamber above orifice, C_d is the coefficient of discharge for the orifice and g is the gravitational constant.

11.6 Urban Drainage Analysis

An urban drainage system is usually made up of connections of several pipes and different types of pits integrated with open channel and surface runoff flows, which makes it a very complex task to be able to analyse the whole system through manual calculations. For the analysis having series of several pipes and pits, some practitioners use a specially developed customised spreadsheet. However, for a more complex sub-catchment, such a spreadsheet should be carefully used, as the possibility of erroneous input data and subsequent wrong analysis is very high. As such, for a whole catchment/sub-catchment drainage analysis, it is recommended that customised modelling tool(s) be used. DRAINS, developed by Watercom Pty Ltd. (www.watercom.com.au), is a widely used sophisticated customised tool for urban drainage system analysis.

In DRAINS, preliminary pits' and pipes' locations are to be delineated based on catchment topography and street locations/alignments. Figure 11.18 shows a sub-catchment, where preliminary pits' locations and pipe alignments are shown. In the figure, small arrows show the direction of flow based on land/road surface slopes. Square blocks (with encircled numbers) are the pits and double dashed lines are the underground pipes connected with the pits. The squiggly arrows represent flows within the pipes. Then each pit is associated with a smaller sub-catchment, from where the flow is likely to enter into that particular pit (some pits may not take any flow from the surface and just act as a junction pit). Such delineations of sub-catchments for all the pits are shown in the Figure 11.19. Catchment data needs to be provided for each catchment; pipe and pit data also need to be provided, design rainfall needs to be selected

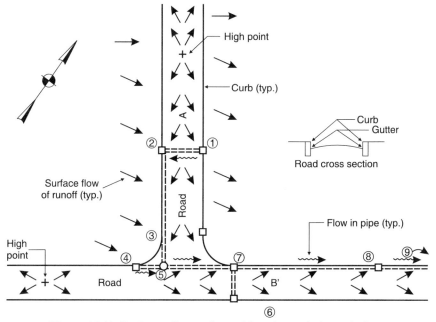

Figure 11.18. Drainage flowpaths and locations of pits and pipes

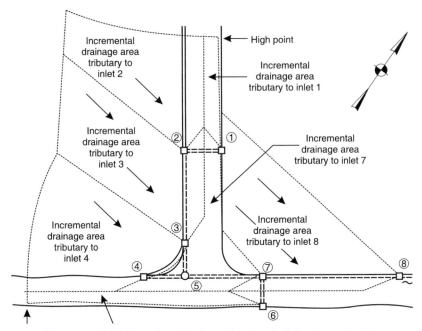

Figure 11.19. Delineation of sub-catchment draining to each pit

from the stored design rainfall database (being an Australian software it stores only Australian design rainfalls, however any rainfall magnitude and pattern from anywhere can be inserted as a synthetic rainfall data). Then, DRAINS simulates the total system based on water level data at the downstream exit pit(s).

The DRAINS model can analyse more hydraulic structures, such as open channel, OSD basin, parallel pipes and different types of pits (however, current pit and pipe databases are based on Australian typical pits and pipes). A detailed description of the DRAINS model and its operation/use can be found from the DRAINS user manual available at: www.watercom.com.au.

11.7 Pit Location Design and Bypass Flow

In an urban drainage system, pits are positioned in strategic locations in order to maximise the entry of overland flow into the pipe system. Sag pits are placed at the lowest point, whereas on-grade pits are placed along the street kerb, which connects the street gutter flow to the stormwater pipe (Figure 11.20). To reduce flooding along the street gutter and to achieve smooth draining of gutter flow into the drainage pipe, it is recommended to maximise the number of on-grade pits. However, due to the cost constraint, an optimum number of pits need to be designed, which requires further analysis. Figure 11.20 shows a schematic of street gutter flow and flows entering into stormwater pipes through pits. In regard to on-grade pit, in reality, the pit does not capture the flow as per its designed capacity. Due to the momentum of water flow towards downslope, some flow bypasses the inlet even if the pit inlet has enough capacity. The steeper the longitudinal slope of the street is, the higher the bypass amount (Figure 11.21). It is to be noted that Figure 11.21 is valid for

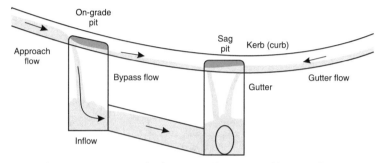

Figure 11.20. Typical schematic of gutter and bypass flows
(*Source*: DRAINS User Manual)

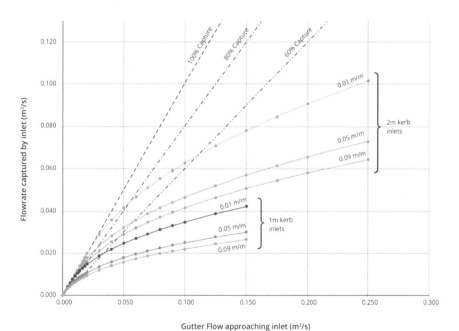

Figure 11.21. Relationship of gutter approach flow and captured flow
(*Source*: Ball et al., 2016)

Color version at the end of the book

Australian standard kerb inlet pits of 1m and 2m wide. For other type of pits such relationship is required to be developed.

Figure 11.22 shows width-discharge relationship curves for a particular gutter configuration (cross-fall, kerb shape) for different longitudinal slopes. Regulatory authority defines a permissible street flow width under a certain ARI/AEP event. Practitioners develop/use curves like Figure 11.22 in order to set allowable flow along the gutter. Then, with the help of Figure 11.21 and catchment (associated with the pits) analysis, the practitioners decide on pit size and/or distances within pits. Figure 11.23 shows a plan of a typical exercise involving land development and subdivision works beside a street, which requires design of pit size and distance. In the figure, the shaded area is the blocks of land beside a street, the dashed line in the middle of the street is the street crown. Arrows depict the slope of the land/street and runoff flow directions. Rectangular solid boxes are the pits.

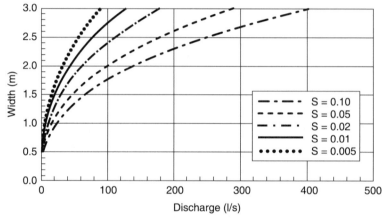

Figure 11.22. Width-discharge relationship for gutter flow

11.8 Overland Flowpath

In urban development, stormwater drainage pipes are underground and often run along the street, and sometime across the house boundaries. The goal is that stormwater passes through the pipes to downstream discharge points (usually open channel or lake) without creating any nuisance to the surrounding properties or users. However, these pipes are designed for a certain ARI/AEP event. For any higher (than designed) ARI/AEP event, such a pipe is likely to overflow and cause flooding to surrounding areas. To manage such overflows during higher events, a special provision should be made so that such overflows do not cause flooding to the neighbouring houses. The standard practice is to provide a design overland flowpath which can contain overflows up to certain ARI/AEP event. Figure 11.24 shows a dead-end (curl-de-sac) of a street that

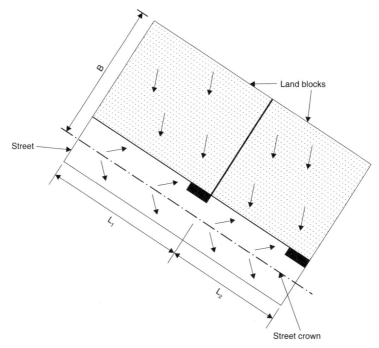

Figure 11.23. A typical example of land subdivision involving pit design

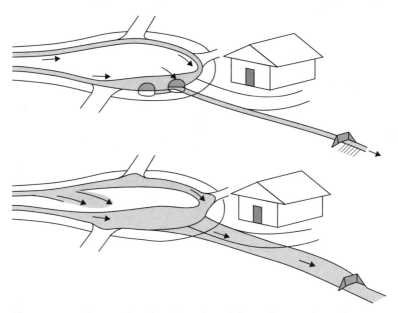

Figure 11.24. Flow within (top figure) and above (bottom figure) a pipe
(*Source*: Pilgrim, 2001)

is carrying runoff and drainage through an underground pipe running beside a house. As long the flows are within the pipe, the neighbouring residents are safe. However, for higher-than-designed flow, the pipe will overflow (lower part of Figure 11.24) and may cause flooding to the neighbouring property, which is a concern for the neighbouring residents. As such, a safe overflow passage/path is required, which will safely drain this overflow to the next point of discharge or open channel/lake. Such an overland flowpath is designed using Manning's equation up to a certain ARI/AEP discharges.

Figure 11.25 shows a photo of such an overland flowpath along the boundaries of houses. A big steel grate is visible (which marks a pit underneath connected with stormwater pipes), the grate opening allows overflow during higher events and a rectangular cut in the side shallow wall, as well as space below fence further downstream, are the components of the overland flowpath which were provided for the safe passage of any overflow above the pipe.

Figure 11.25. An overland flowpath in between property boundaries

Worked Example 1

In an urban development, two pit distances must be designed (refer to Figure 11.24). Both the pits are associated with a sub-catchment (including half street width) having a width of 50 m. The street gutter is allowed to carry 0.12 m^3/s of stormwater. Determine distances L_1 and L_2, for a design rainfall of 80 mm/hr considering a runoff co-efficient (C) of 0.90. Also, assume that the first (from left) pit has a capture rate (capability) of 70%.

Solution:

$I = 80$ mm/hr, $C = 0.90$

Allowable flowrate (Q) = 0.12 m^3/s

Using the runoff equation for the first (from left) block, $Q = C*I*A/360 = 0.90*80*(50*L_1/10,000)/360 = 0.12$ (allowable flowrate), which yields, $L_1 = 120$ m.

Now, for the next pit, bypass flow is 30%, as the flow capture rate is 70% by the first pit. So, the next pit would be able to take from an area which is only 70% of the first catchment, which means the $L_2 = 0.7*L_1 = 84$ m.

Worked Example 2

Figure 11.24 shows a carpark draining to a street with two pits. Both the pits are associated with separate sub-catchments (including half the street width) that have a width of 50 m. Considering the allowable street gutter flow capacity, the distance, L_1 was calculated to be 80 m. Calculate, the allowable street gutter flow in m^3/s. The design rainfall considered was 100 mm/hr and an impervious surface with runoff co-efficient $C = 0.90$. The longitudinal slope of the street is 0.01 m/m and the pit is a 2 m wide kerb inlet. Using Figure 11.22, determine the amount of flow bypass from the first pit (from left).

Solution:

$I = 100$ mm/hr, $C = 0.9$

Using the runoff equation for the first (from left) block, $Q = C*I*A/360 = 0.9*100*(50*80/10,000)/360 = 0.1$ m^3/s.

So, the allowable gutter flow is 0.1 m^3/s.

From the Figure 11.22, for a 0.01 m/m slope and a 2 m kerb inlet pit, with the approach flow of 0.1 m^3/s, the capture flow would be 0.065 m^3/s.

So, the bypass flow = 0.1 – 0.065 = 0.035 m^3/s.

Worked Example 3

For a subdivision pit design task, the authority recommends pits to be designed for a rainfall intensity of 80 mm/hr using a runoff co-efficient (C) of "0.85". The adjacent street's longitudinal slope is 5% and authority allowable street flow width is 2.0 m for the mentioned design rainfall. Half of the street drains to the pits adjacent to the blocks and the pits are to be

2 m wide. Determine the pit distances L_1 and L_2 (refer to Figure 11.24), considering that the total width of the land blocks plus half of the street (B) is 40 m. Use the width-discharge relationship shown in the Figure 11.23.

Solution:

I = 80 mm/hr, C = 0.85

From the Figure 11.23, for a 5% slope to maintain a maximum flow width of 2.0 m the allowable flowrate (Q) = 100 l/s = 0.1 m³/s.

Using the runoff equation for the first (from left) block, $Q = C*I*A/360 =$ 0.85*80*(40*L_1/10,000)/360 = 0.1 (allowable flowrate), which yields, L_1 = 132.35 m.

Now, for the next pit, the captured flow by the first pit needs to be evaluated. From the Figure 11.22, for a 2 m wide pit along a 5% slope street, if approach flow is 0.1 m³/s, the captured flow would be 0.0475 m³/s. So, the bypass flow = 0.1 – 0.0475 = 0.0525 m³/s. So, the next pit is allowed to take only 0.1 – 0.0525 = 0.0475 m³/s from the adjacent block.

Using the runoff equation for this flow, 0.0475 = 0.85*80*(40*L_2/10,000)/360, which yields, L_2 = 0.85*80*(40*/10,000)/360 = 62.87 m

Worked Example 4

Figure 11.24 shows a carpark draining into a street with two pits. Both the pits are associated with a sub-catchment (including half street width). The street gutter is allowed to carry 0.1 m³/s of stormwater. Distances L_1 and L_2 were calculated to be 60 m and 36 m, respectively, for a design rainfall of 100 mm/hr, considering an impervious surface (runoff co-efficient C = 0.9). Using Figure 11.22, determine the maximum allowable longitudinal slope for this road, if 2 m wide kerb inlets are to be used.

Solution:

For the same width (50 m), ratio of L_2 to L_1 = 36/60 = 0.60
So, the capture flow rate into the pit = 0.60*0.1 = 0.06 m³/s

From the Figure 11.22, for an approach flowrate of 0.1 m³/s and capture flowrate of 0.06 m³/s, for a 2 m kerb inlet the longitudinal slope should not be more than 0.02 m/m (approximate interpolation between the curves for 0.01 and 0.05 m/m).

Worked Example 5

Figure 11.24 shows a carpark draining into a street with two pits. Both the pits are associated with a sub-catchment (including half street width) that has a width of 50 m. Based on allowable flow along the street gutter, two pit distances were calculated as L_1 = 100 m and L_2 = 25 m. For the analysis, it was considered that the carpark and associated street catchment is 100% impervious; design rainfall intensity = 72 mm/hr. Using the Figures 11.22 and 11.23, determine the longitudinal slope of the street. Also, calculate the allowable flow width on the street using the attached chart. Consider that the pits are 1 m wide kerb inlet pits.

Solution:

Using the runoff equation for the first (from left) block, $Q = C*I*A/360 = 1.0*72*(50*100/10,000)/360 = 0.1$ m³/s (allowable flowrate).

As L_2 is 25% of L_1, the capture flow is $0.1*0.25 = 0.025$ m³/s

From Figure 11.22, for the approach flow of 0.1 m³/s and capture flow of 0.025 m³/s, the longitudinal slope is approximately 0.05 m/m.

From Figure 11.23, for a flow of 0.1 m³/s (= 100 l/s) and slope 0.05 m/m, the allowable flow width is 2.0 m.

Worked Example 6

The figure below shows two pits and a pipe section of a continuous stormwater pipe system. All the (pit invert and obverts) levels are provided. A water flow of 0.15 m³/s is coming from upstream. The pipe is 50 m in length and has a diameter of 300 mm. The water level at Pit 2 is 41.1 m. Determine the water level at Pit 1. Assume that pit loss co-efficient $K_u = 0.5$ and friction loss co-efficient $f = 0.01$.

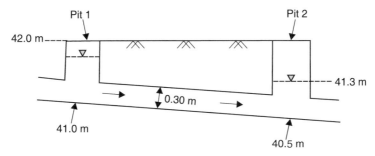

Solution:

Pipe diameter, $D = 300$ mm = 0.30 m

Velocity of the flow within the pipe, $V = Q/A = 0.15/(\pi/4*0.3^2) = 2.12$ m/s

Pit head loss = $K_u*V^2/(2\ g) = 0.115$ m

Using Equation 11.8, head loss due to pipe friction = $f*(L/D)*V^2/(2\ g)$

$= 0.01*(50/0.3)*2.12^2/(2*9.81) = 0.382$ m

Using Equation 11.10, water level (static head) at 'Pit 1' = 41.3 + 0.382 + 0.115 = 41.80 m

References

Ball, J., Babister, M., Nathan, R., Weeks, W., Weinmann, E., Retallick, M., Testoni, I. (Editors) (2016). Australian Rainfall and Runoff: A Guide to Flood Estimation. © Commonwealth of Australia (Geoscience Australia).

Colebrook, C.F. (1939). Turbulent flow in pipes, with particular reference to the transition region between the smooth and rough pipe laws. Journal of the Institution of Civil Engineers 11(4): 133–156. doi:10.1680/ijoti.1939.13150.

Pilgrim, D.H. (2001). Australian Rainfall and Runoff: A Guide to Flood Estimation, Vol. 1. Institution of Engineers Australia, Barton, ACT. ISBN: 0858257440.

Water Conservation and Recycling

12.1 Introduction

With increasing population and changing climate regime, water supply systems in many cities of the world are under stress. For the arid countries surrounded by ocean, desalination is the only option in many cases. However, installation and maintenance of a desalination plant is very expensive and not affordable for all countries. Countries with adequate surface and groundwater have been using water lavishly. However, as population and urbanisation are rapidly increasing, so too is the demand for potable water, while on the other hand, due to impact of climate change, supply (mainly rainfall) is becoming uncertain/non-uniform in many parts of the world. To tackle this problem, water authorities in many cities are adopting several measures, including demand management and identifying alternative water sources, such as stormwater harvesting, greywater and wastewater reuse and desalination. To achieve a significant alleviation on current excessive water uses, all the possible ways need to be implemented; starting from individual habit/consumption, through to household, local and national demands. People often think that individual optimum use of water is enough to reduce total water use, however, total water use also comprises some other components, such as irrigation water use and industrial water use. As such, to account for a country's total water use, a new term named "Water Footprint" has evolved, which will be discussed in the following section.

12.2 Water Footprint

On an individual level 'water footprint' is simply the amount of water used at a personal level. However, on a national level, the 'water footprint' is equal to the total use of domestic water resources, minus the virtual water export flows, plus the virtual water import flows. The virtual water as the amount of water that is embedded in food or other products needed for its production. The following are some virtual water amounts embedded in different foods/products (note: These amounts will vary from country to country, depending on the system/process in place):

- 1 cup of coffee: 140 litres
- 1 litre milk: 800 litres
- 1 kg wheat: 1,100 litres
- 1 kg rice: 2,300 litres
- 1 kg beef: 22,000 litres
- A cotton T-shirt: 2,000 litres
- Paper: 23,400 litres/tonne
- One tonne vehicle: 47,600 litres

Figure 12.1 schematically shows the different components of total water footprint.

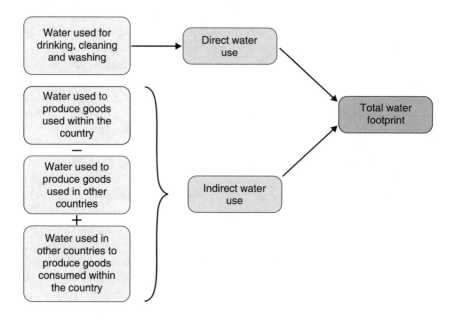

Figure 12.1. Components of total water footprint

In general, total consumption of a country/region is the total amount of water used/withdrawn for domestic, commercial/industrial and agricultural purposes. However, this total amount is often divided by the number of residents in order to get an average per capita consumption. Figure 12.2 shows the per capita water withdrawal of different countries.

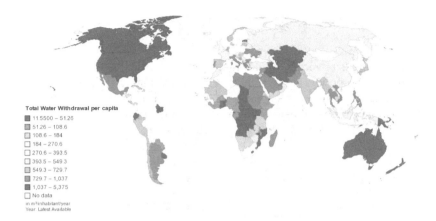

Total Water Withdrawal per capita
- 11.5500 – 51.26
- 51.26 – 108.6
- 108.6 – 184
- 184 – 270.6
- 270.6 – 393.5
- 393.5 – 549.3
- 549.3 – 729.7
- 729.7 – 1,037
- 1,037 – 5,375
- No data

in m³/inhabitant/year
Year: Latest Available

Figure 12.2. Total per capita water withdrawal of different countries
(*Source*: http://chartsbin.com/view/1455)

Color version at the end of the book

12.3 Sustainable Water Fixtures

To achieve a sustainable water future, the first step is to start with individual and/or household level. Most of the affluent countries have been using too much water, even for their individual needs in addition to the higher level of commercial/industrial needs. Government authorities around the world have been campaigning and encouraging lower water consumption at an individual level. To facilitate this, they have been promoting the use of different sustainable water fixtures (i.e. tap, shower head) that use less water (i.e. restrict water flow). In addition, manufacturers are providing different water efficient appliances (i.e. dishwasher, washing machine). Many of these products now come with 'energy' and 'water' star ratings; higher the star ratings, better it is for the environment and sustainability.

12.4 Stormwater Harvesting

Among all the alternative water sources, stormwater harvesting perhaps has received the highest level of attention, as it is cheap, easy to install and does not have any major safety concern. Also, it is easy to catch/collect rainwater and house roofs are generally used for the collection. Moreover, it requires little treatment and maintenance. However, it is dependent on the regions where rain occurs with a significant amount. In some parts of the world (i.e. arid areas), there is no or an insignificant amount of rain. In many remote areas where significant rain occurs, stormwater harvesting has been in practice for many centuries as there are no other suitable/significant source of water. Many of these communities consume roof-collected rainwater, after some basic treatment. Although, in the past, developed nations (especially in the city areas) did not bother to use rainwater tanks, however, with recent impacts of climate change, as well as ever-increasing water demand, water authorities in the developed cities are also encouraging the installation of rainwater tanks and use of rainwater for non-potable purposes. Many government authorities have even been paying a subsidy/bonus for rainwater tank installations.

In an urban setting, where population and house density is high, the space required for a rainwater tank is an issue. In general, a bigger sized tank is expected to save more water i.e. the greater the tank, the size better it is. However, as water savings depend on some other factors (roof size, water consumption) as well, depending on those factors, there is a threshold size, beyond which a bigger tank does not provide higher water savings. Figure 12.3 shows typical relationships of annual water savings with tank size and roof area for a rainwater demand of 300 L/day in an Australian city. Again, such a relationship will vary with the rainwater demand; in general, the higher the rainwater demand, the higher the annual water savings. Several studies have presented detailed water savings scenarios for different demands, as well as roof and tank sizes for different Australian cities, i.e. Sydney (Rahman et al., 2012), Melbourne (Imteaz et al., 2016b), Canberra (Imteaz et al., 2014) and Adelaide (Imteaz et al., 2016a).

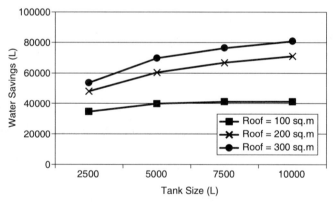

Figure 12.3. Typical relationship between water savings, roof and tank sizes

From the figure, it is clear that for roof sizes of 100 and 200 m^2, a tank beyond the size of 10,000 L is not expected to provide any additional water savings for the selected demand. Even for a roof of 300 m^2, the threshold tank size is very close or slightly above 10,000 L. It is to be noted that connecting a rainwater tank with a roof size of 300 m^2 is a huge job and a roof of 300 m^2 is not very common for a single house. Deciding an optimum rainwater tank size is crucial, as in addition to the factors of roof size and rainwater demand, it also depends on weather conditions.

Especially for the regions where inter-annual climatic variability is high, in a dry year as amount of rainfalls will be low, even a small sized tank is not likely to overflow. Whereas, in a wet year, as the amount of rainfalls will be high and frequent, even a big tank is likely to overflow. Imteaz et al. (2017b) have presented a detailed investigation on climatic and spatial variability of rainwater tank outcomes for an Australian city, Sydney.

In spite of the benefit of an augmented water supply and several government initiatives, in many cases, a general reluctance is evident in installing rainwater tanks, as individual owners/residents mainly look for their own monetary benefit. In most of the developed countries, the installation and maintenance of a rainwater tank is expensive, while the price of potable water supplied to the houses is cheap. In such cases, the payback period of household's individual rainwater tank installation and maintenance costs become very long. Payback period is the period (usually in year) required to get an accumulated benefit equals to (or above) the total installation and maintenance costs during that period. To diminish such reluctance on installing rainwater tanks, many government authorities offer a certain benefit/subsidy towards the installation/purchase cost of a rainwater tank. Another indirect option to reduce the payback period, is to increase the potable water price. Figure 12.4 shows a typical relationship of payback period with the water price increase for a large rainwater tank in the city of Melbourne (Australia) and Imteaz et al. (2011) have provided a detailed investigation on payback period and water price considering large water tanks installed within the Swinburne University of Technology campus at Hawthorn, Melbourne. From this typical example, it is seen that the payback period can significantly drop down from a long period of "44" years to a short period of "15" years with the increase of potable water price. Nonetheless, residents do not like

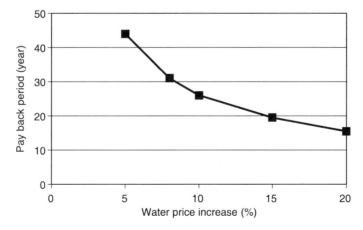

Figure 12.4. Typical relationship of payback period with water price

such a price hike. However, when the government's expenditure is the concern, then optimum combination(s) of potable water price hike and/or reasonable subsidy/incentive towards rainwater tank installations need to be adopted. Imteaz and Moniruzzaman (2018) have presented a detailed study on reasonable government rebates considering several case studies for Sydney, Australia. Imteaz et al. (2017a) have developed a computer tool, eTank, which analyses rainwater tank outcomes including payback period for any combination of input data.

12.5 Greywater Recycling

Greywater is termed as the water which drains out from our bathroom, kitchen, hand basins and laundry. In general, water from these usages is not highly contaminated and on the other hand some other usages (i.e. toilet flushing, garden irrigation) do not require high quality water. As such, greywater can be reused for some lower-level usages. In reality, wastewater from laundry, kitchen and bathroom is more suitable to the garden, as this water is likely to be enriched with higher level nutrients, which are good for plants. However, some preliminary treatments are required before greywater can be reused. There are commercial treatment devices, which can do such preliminary level treatments and can be installed within a house basement or garage. Also, additional plumbing connections are required in order to connect treated water with the intended usages (i.e. toilet, garden). For such treatment and additional plumbing requirements, recycling greywater becomes an expensive option in some cases. For a single household where demand of greywater is not high, this option is not likely to be economically feasible, as it will only be saving potable water from a single household for few specific usages (which is usually not high). For a multistorey building having many flats, this option may turn out to be economically feasible, as demand of greywater from many flats is likely to be high. Also, as many flats will be sharing the installation and maintenance costs of the treatment device(s), costs per household will not be high. Most multistorey buildings do not have rooftop gardens, so there is no irrigation requirement. Therefore, for multistorey buildings, the use of greywater is only for the toilet flushing. As such, the amount of greywater generated is usually higher than the amount of greywater needed, so it is not wise to collect greywater from all the flats, but it is recommended to collect greywater from some upper level flats and supply it to all the levels. Through such a partial collection option, the overall cost can be minimized. Figure 12. 5 shows a schematic diagram of greywater collection and supply to the toilets for a multistorey building (Shanableh et al., 2012).

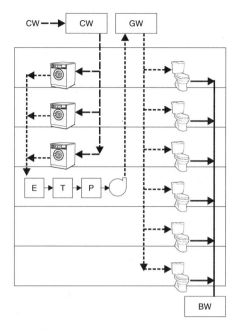

Figure 12.5. Schematic diagram of greywater recycling from multistorey building (CW: Clean water, GW: Greywater, BW: Blackwater, E: Equalizer, T: Treatment, P: Pump)

12.6 Centralised Recycled Water

Since supplying and managing wastewater for reuse and recycling purposes require very high-level treatments, monitoring and quality control measures, often many private owners do not prefer to adopt it. To maintain proper quality control checks, government authorities may adopt such a practice by providing centralised wastewater collection, treatment and resupply. Some cities had to adopt this when affected by severe drought for a prolonged period, even though many residents were not ready to accept such recycled water. The reality is that nature has been gradually recycling all the wastewater through a longer process, which is called natural hydrologic cycle discussed in the Chapter 1. The water that we are drinking now (or using for other foods) might have been wastewater produced by someone else (or by some city/village) hundreds/thousands of years ago. Considering this issue of residents' acceptability, many new suburbs in Australia adopted a lower level of localised treatment of wastewater and recycle it for irrigation/outdoor or toilet flushing purposes only. Figure 12.6 shows a photo of such recycled water supply with a warning sign/message, beside a normal water supply tap.

Figure 12.6. Photo of recycled water supply pipe next to a normal
water supply pipe

References

Imteaz, M.A. and Moniruzzaman, M. (2018). Spatial variability of reasonable
government rebates for rainwater tank installations: A case study for Sydney.
Resources, Conservation & Recycling 133: 112–119.

Imteaz, M.A., Karki, R., Shamseldin, A. and Matos, C. (2017a). eTank and
contemporary online tools for rainwater tank outcomes analysis. International
Journal of Computer Aided Engineering and Technology 9(3): 372–384.

Imteaz, M.A., Moniruzzaman, M. and Karim, M.R. (2017b). Rainwater tank analysis
tools, climatic and spatial variability: A case study for Sydney. International
Journal of Water 11(3): 251–265.

Imteaz, M.A., Paudel, U., Matos, C. and Ahsan, A. (2016a). Generalized equations
for rainwater tank outcomes under different climatic conditions: A case study
for Adelaide. International Journal of Water 10(4): 301–314.

Imteaz, M.A., Sagar, K., Santos, C. and Ahsan, A. (2016b). Climatic and spatial
variations of potential rainwater savings for Melbourne. International Journal
of Hydrology Science and Technology 6(1): 45–61.

Imteaz, M.A., Matos, C. and Shanableh, A. (2014). Impacts of climatic variability
on rainwater tank outcomes for an inland city, Canberra. International Journal
of Hydrology Science and Technology 4(3): 177–191.

Rahman, A., Keane, J. and Imteaz, M.A. (2012). Rainwater harvesting in greater
Sydney: Water savings, reliability and economic benefits. Resources,
Conservation & Recycling 61: 16–21.

Imteaz, M.A., Shanableh, A., Rahman, A. and Ahsan, A. (2011). Optimisation of rainwater tank design from large roofs: A case study in Melbourne, Australia. Resources, Conservation & Recycling 55(11): 1022–1029.

Shanableh, A., Imteaz, M.A., Merabtene, T. and Ahsan, A. (2012). A Framework for reducing water demand in multi-storey and detached dwellings in the United Arab Emirates. 7th International Conference on Water Sensitive Urban Design, Melbourne, February.

Water Sensitive Urban Design

13.1 Introduction

Efficient drainage systems in urban areas were introduced for the safe passage of urban runoff to the downstream stream/water body. Since they involve huge capital costs, developed countries were obviously ahead in implementing such infrastructures (pipes, pits, culverts) with proper planning and design. Such efficient drainage systems were thought to be a blessing to urban residents. However, with the rapid urbanisation having such an efficient drainage system, the problem of concentrated runoff/ flooding shifted from the catchment source to a downstream location with even higher magnitude as several such drainage systems discharge in a single location. Urbanisation not only increases the flood peak, but also reduces the time to peak, the consequences of which often become worse. Due to the shorter time to peak, residents do not get enough time to evacuate (or remove their belongings) before a large flood event occurs in the locality. Moreover, in the past, urban stormwater runoff was considered as having minimal contaminants. However, more and more urbanisation and increased population, as well as increased anthropogenic activities using different chemicals (detergent, car oil, insecticide, pesticide etc.), is resulting in heavily polluted urban runoff. The sum of all these polluted runoffs from different catchments discharging in a single location (bay/ lake) is no longer insignificant and requires attention. Among controlling measures, the most useful is in source control or control while conveying. As such, contemporary city developers prefer a drainage system that uses natural means, which will retard and reduce the flood. A natural surface will convey runoff, while reducing its volume (through infiltration) and lengthening the time to peak (through providing rough natural surfaces with grass/plants/shrubs). Moreover, such a natural surface will also provide some pollution treatment, as filtration occurs when polluted water passes underground through the natural surface(s).

Designing a city with such feature(s), that use or mimic natural processes which increases the infiltration and evapotranspiration and eventually reduce the runoff volume, retard the runoff flow and at the same time will treat the runoff to some extent, is called Water Sensitive Urban Design (WSUD). However, many other countries are also implementing such measures with different names; i.e. in the USA, Canada and New Zealand it is called Low Impact Development (LID) or Green Infrastructure (GI). In the UK, these are called Sustainable Drainage Systems (SuDS), whereas in Malaysia, these are called Sustainable Urban Drainage Systems (SUDS). In the following sections, some of these features are described in detail.

13.2 Grass Swale

Grass swale, or "vegetated swale", provides a very basic level of treatment at the source of the pollutants generation by providing grass integrated with natural surroundings. For example, instead of providing concrete/impervious road-side drainage (Figure 13.1), it is recommended to provide drainage with a natural surface and grass (Figure 13.2). Such grass traps some coarser sediments, through which some nutrients are also being trapped as nutrients are usually attached with the sediments.

There are some design criteria recommended to be used while designing such an overland drainage path. Typically, Manning's equation

Figure 13.1. Roadside drainage using impervious surface

Figure 13.2. Roadside drainage using natural (grass) surface

is used to size the swale, given the site conditions. This calculation is sensitive to the selection of Manning's *n* and this should vary according to flow depth (as it decreases significantly once flow depths exceed vegetation height). In fact, it varies with flow depth, channel dimensions and the vegetation type.

For constructed grass swale systems, the 'n' values are recommended to be between 0.15 and 0.4 for flow depths shallower than the vegetation height (preferable for treatment) and can be significantly lower (e.g. 0.03) for flows with greater depth than the vegetation (however, this can vary greatly with channel slope and cross-section configuration). A critical design parameter is the design velocity along the channel, which needs to be satisfied considering potential damage of grass and/or channel surface. The typical criteria used in Australia are:

- Velocity along the channel should be checked for the following conditions
  - Velocity less than 0.5 m/s for minor storm (e.g. 5-year ARI) discharges
  - Velocity less than 1.0 m/s for major storm (e.g. 100-year ARI) discharges
- The flow depths and velocities need to be acceptable from a public risk perspective

In Australia, the following standard is used for the design flow rate: Velocity (m/s) × depth (m) < 0.4 m^2/s

In fact, this is the simplest form of feature among all the WSUD features, without requiring much maintenance. As such, it is recommended that this should be widely used in our urban settings, wherever possible. Details of the grass swale design procedures can be found from Melbourne Water (2005).

13.3 Sand Filter

The next higher level of WSUD feature is 'sand filter', which provides basic filtration of stormwater through sand as filter media, before it reaches to a nearby piped drainage system. Figure 13.3 shows a typical sand filter built to treat stormwater draining from a car park. In the figure only top gravel layer is seen, however the sand layer is placed at the bottom of the gravel layer and, at the bottom of the sand layer, perforated drainage pipes are installed in order to collect treated stormwater after filtration.

Figure 13.3. A sand filter (underneath the gravel layer) in a car park

In the figure, the plant is mainly for aesthetic purpose, not really a bioretention plant (will be discussed in the following section). In a bioretention system, typical shrubs are densely placed, as shown in Figure 13.6.

$$Q_{max} = k * A * \left(\frac{h_{max} + d}{d} \right) \tag{13.1}$$

where Q_{max} is the maximum infiltration rate (m³/s), k is the Hydraulic conductivity of the soil filter (m/s), A is Surface area of the sand filter (m²), h_{max} is the depth of pondage above the sand filter (m), d is the depth of filter media (m).

The following equation is used to design the size of the pipe which will carry the treated stormwater to the main drainage system:

$$Q_{pipe} = C * A_{pipe} * \sqrt{2gh} \tag{13.2}$$

where Q_{pipe} is the discharge capacity of each pipe (m³/s), A_{pipe} is the area of each pipe (m²), C is the orifice coefficient (~0.6) and h is the head above slotted pipe (m).

In regard to perforations in the pipe, the discharge capacities of the holes are calculated using the following equation:

$$Q_{perforation} = C * BF * A_{perforation} * \sqrt{2gh} \qquad (13.3)$$

where $Q_{perforation}$ is the discharge capacity of each hole (m³/s), $A_{perforation}$ is the area of each slot/perforation (m²), h is the head above slotted pipe (m), C is the orifice coefficient (~0.6) and BF is the blockage factor due to granules (~0.5).

For any such system, a bypass passage needs to be provided in order to bypass the overflow which is in excess to designed flow capacity. This is usually provided by installing a weir at a certain designed level, which will start bypassing the excess flow to the downstream drainage system without filtration/treatment. The weir equation in the following format is used to design the weir length and the head above the weir:

$$Q_{weir} = C_w * L * H^{1.5} \qquad (13.4)$$

where Q_{weir} is the design discharge over the weir in m³/s (usually taken as 100 years or 50 years flow), C_w is the weir coefficient (~1.7), L is the length of the weir (m) and H is the design head above the weir (m).

In many occasions where space is available, an initial sedimentation chamber is provided before the sand filter in order to trap the heavier sediments and debris so as to avoid blocking of the main filtration system. Figures 13.4 and 13.5 show a typical sand filter's plan and cross-sectional views, respectively. Details of the sand filter design procedures can be found from Melbourne Water (2005).

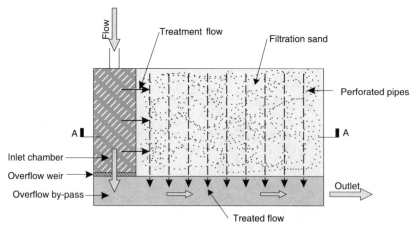

Figure 13.4. Plan view of a typical sand filter with initial sediment chamber
(*Source*: Melbourne Water, 2005)

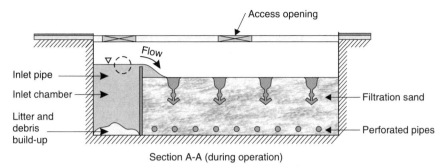

Figure 13.5. Cross-sectional view of a typical sand filter
(*Source*: Melbourne Water, 2005)

A sand filter is primarily used to trap sediments from the stormwater. However, as some nutrients are attached to the sediment surface, trapping sediments also helps to indirectly trap nutrients (from stormwater) to some extent.

13.4 Bioretention System

The bioretention system is basically a sand filter with the added feature of having plants/shrubs in it, as shown in Figure 13.6. Due to the presence of plants/shrubs and their roots, a bioretention system is expected to trap more nutrients (phosphorus and/or nitrogen), in comparison to a sand filter, through some biological actions of the roots. Moreover, these plants/shrubs add aesthetics to the urban environment.

The hydraulic capacities and calculations for bioretention systems are the same as for sand filters, the only exception might be that, due to the presence of vegetation roots, the hydraulic conductivity value of the filter media

Figure 13.6. A bioretention system in between two road strips

is expected to be higher. In regard to treatment capabilities, there have been several field investigations carried out around the world. Imteaz et al. (2013) summarised some of those investigations and compared the outcomes with the MUSIC (Modelling Urban Stormwater Improvement

Conceptualisation) simulated results. MUSIC, a powerful modelling tool, was developed by eWater (https://ewater.org.au/products/music/). MUSIC can simulate generation of pollutants from different types of catchments, as well as pollutants' treatment efficiencies through different types of WSUD measures, including grass swale, sand filter, bioretention system, sedimentation basin, detention basin, rainwater tank and wetland. Details of the bioretention system design procedures can be found from Melbourne Water (2005)

13.5 Porous Pavement

Porous/permeable pavements are contemporary urban pavements mainly used for lightweight traffic movements, such as pedestrians and bicycles. They are also used in parking lots, emergency access lanes, service roads, sidewalks and driveways. Porous pavements are usually constructed with any of the four basic material types: i) porous asphalt, ii) porous concrete, iii) interlocking paver blocks and iv) plastic grids, preferably with underlying sand/filter media. Such pavements reduce the road runoff, allowing some stormwater to infiltrate through the voids of the pavement and finally through the sand filter. As the basic function is filtration, it will trap sediments which, in turn, will also trap some nutrients, as discussed earlier. A typical demonstration of porous pavement is shown in Figure 13.7. It is to be noted that the porous pavement is not suitable for heavy vehicular movements.

Figure 13.7. A typical demonstration of porous pavement
(https://commons.wikimedia.org/wiki/File:Permeable_paver_demonstration.jpg)

As porous pavement is basically a filtration measure, predominantly trapping sediments from stormwater, the prime treatment it provides is the removal of sediments. However, as mentioned earlier, trapping sediments also traps some nutrients which are attached to the trapped sediments. Details of the porous pavement design procedures can be found from Melbourne Water (2005)

13.6 Sedimentation Basin

Sedimentation basins have been in use around the world for many decades. Although, in the recent past, due to space limitation and demand for more urban housing, many urban authorities have been failing to provide them. Recently, with rapid urban developments and subsequent generation of more and more urban pollutants, the need to provide sedimentation basins has become more prominent than before. It applies the very basic technique of settling sediments by reducing water flow/velocity or by providing enough retention time for the sediment to settle. Details of the sedimentation mechanism and equations are described in Chapter 9. Figure 13.8 shows a typical cross-section layout of a sediment basin.

For the sizing of sedimentation basins a compromise is often required; i.e. to capture a finer sediment either a large surface area has to be provided or flowrate of treatment should be very low. In practice in an urban setting, to get a bigger area is often very expensive and difficult. On

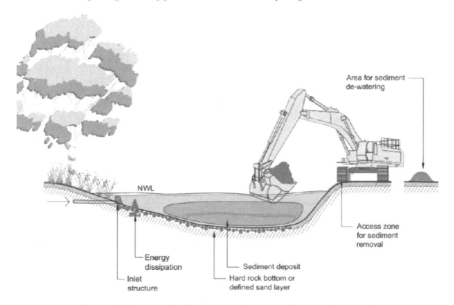

Figure 13.8. A typical cross-section of sedimentation basin
(*Source*: Melbourne Water, 2005)

the other hand amount of stormwater needed to be treated is increasing day by day. As such, often a realistic and larger target sediment size for capture is considered. For a larger sedimentation basin, the future maintenance/cleaning facility/provision is an important parameter, as it is very difficult to clean/remove accumulated sediments from a very large sedimentation basin, as a significant amount of sediments might have been deposited in the areas which are out of reach of common excavation machineries. For some other more sophisticated treatment facilities (i.e. sand filter, wetland, etc.), providing a sedimentation basin prior to the (i.e. upstream) sophisticated system is recommended, as it will reduce the maintenance burden for the downstream sophisticated system.

Equations for particle settling velocity described in Chapter 9 can be used for the calculations of particle settlements in a particular sediment basin. However, to achieve quick decision making, practitioners often use some recommended settlement velocities for different sized particles, considering sediment settlements in normal water temperature (20° C). Table 13.1 shows some recommended values of different types of sediments with different diameters.

Table 13.1. Recommended settling velocities for different types of sediments

Classification of particle size	Particle diameter (μm)	Settling velocity (mm/s)
Very coarse sand	2000	200
Coarse sand	1000	100
Medium sand	500	53
Fine sand	250	26
Very fine sand	125	11
Coarse silt	62	2.3
Medium silt	31	0.66
Fine silt	16	0.18
Very fine silt	8	0.04
Clay	4	0.011

For the sedimentation basins, as well as for wetlands (discussed in the following section), provisions should be made for proper outflow structure(s). For this purpose, either a weir-like structure (Figure 13.9) or an orifice-embedded structure, or a combination of both (Figure 13.10), are used.

For the calculations of such an outflow structure's capacity, the standard weir and orifice equations are used. For the weir outflow, the design parameter is the required perimeter of the weir for a certain design

Figure 13.9. A photo of weir outflow from a sedimentation pond
(*Source*: Melbourne Water, 2005)

Figure 13.10. A photo of combined weir and orifice outflow from a pond
(*Source*: Melbourne Water, 2005)

flow and design depth, which can be determined using the following equation:

$$P = \frac{Q_{des}}{B * C_w * H^{1.5}} \qquad (13.5)$$

where P is the required perimeter of the outlet pit (m), C_w is the weir coefficient (~1.7), H is the depth of water above the crest of the outlet pit (m), Q_{des} is the design discharge (m^3/s) and B is the blockage factor (~0.5).

For an orifice embedded outlet structure, the following can be used:

$$A_o = \frac{Q_{des}}{B * C_d * \sqrt{2gH_o}}$$

(13.6)

where A_o is the required area of the orifice (m^2), Q_{des} is the design discharge (m^3/s), B is the blockage factor (~0.5), C_d is the orifice discharge coefficient (~0.6) and H_o is the depth of water above centre of orifice (m).

For the higher trapping of sediments, it is better that the flow is contained within the basin for a longer period so that the sediments get enough time to settle at the bottom of the basin before it exits through the outlet. However, in urban settings, due to space limitation, it is often very difficult to provide a longer travel path for the moving sediments. As such, in order to achieve a slower flow (i.e. longer flow path), a zig-zag fashioned sediment basin is often provided. Details of the sedimentation basin design procedures can be found from Melbourne Water (2005).

13.7 Wetland

A wetland is typically provided in order to treat pollutants from the stormwater/wastewater in addition to trapping sediments. As wetland favours biogeochemical processes, there should be some plants in the wetland to support these processes. In simple words, a sedimentation basin is a wetland without plants (i.e. without having any biogeochemical processes). A wetland has additional features (i.e. plants), so it is difficult

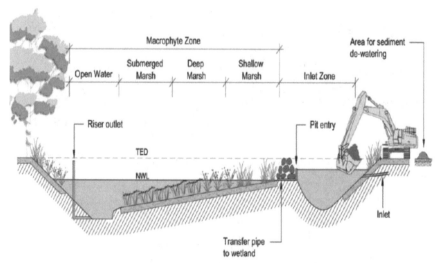

Figure 13.11. A typical cross-section of a wetland
(*Source*: Melbourne Water, 2005)

to remove sediments from such a facility, therefore, it is recommended to provide a sediment basin before the inlet of a wetland in order to trap sediments before the stormwater enters into the wetland. Such a sediment basin should have easy access for sediment removal machineries/ vehicles. Figure 13.11 shows a typical cross-section of wetland with a sediment basin before the wetland inlet. Also, Figure 13.12 shows a photo of a wetland with plants inside it.

Figure 13.12. Photo of a wetland

As wetland provides biochemical treatments through its plants, it is expected to treat nutrients from the water, in addition to trapping sediments (and associated nutrients attached with the sediments) through the sedimentation process. Furthermore, it also provides aesthetic benefits to the residents, as well as providing a habitat for aquatic animals and birds. Details of the wetland design procedures can be found in Melbourne Water (2005).

13.8 Riparian Vegetation

The traditional practice of conveying higher volumes of stormwater (i.e. at the catchment outlet) was to provide a concrete-lined channel with the aim of swift drainage of generated runoff from upstream of a catchment to some downstream locations. However, with the increase of such efficient channels, flooding problems were shifted to some downstream locations, causing recurrent flooding with a shorter time to peak (since with the lined channels, stormwater can travel very quickly to the downstream locations). In many cities, flooding with shorter time to peak has created havoc and hindrance to adequate evacuation processes and to the authorities and/or residents. Moreover, such a lined channel does not provide any aesthetic or environmental values. As such, the WSUD concept recommends that a natural channel with native vegetation be provided for the conveyance of stormwater. Such vegetation along the channel is called riparian vegetation, as shown in Figure 13.13.

Figure 13.13. A photo of a stormwater channel with riparian vegetation

In addition to providing aesthetic values, such riparian vegetation consumes nutrients from the stormwater. Also, as natural soil surface is used, it reduces the runoff volume through infiltration of runoff. Moreover, roots of such vegetation stabilise the channel surface and bottom, protecting them from erosion. The major drawback of providing such riparian vegetation is that it contributes to higher Manning's n value (i.e. roughness), which contributes to higher flood level for the same runoff/discharge.

13.9 Rainwater Tank

The prime purpose of the rainwater tanks is to store the natural downpour of water which would have been wasted to the drainage pipes/channels and to use the stored water to augment the supply of water to household residents. Rainwater tanks have been in use for many centuries, especially for remote communities where other sources of water are scarce. In modern society, areas where people live in single house blocks with gardens, rainwater collected through a rainwater tank is recommended for use in garden irrigation, as the plants in the garden do not need high quality potable water for their survival. Very recently, in order to achieve a more sustainable water use, it is recommended (for regions where rainfalls are significant) that collected rainwater to be used for toilet flushing and/or laundry. All these uses are aimed at reducing the potable water demand from the main water supply. Recently, it has been unearthed that, in addition to providing a significant water saving benefit, rainwater tanks also help to trap sediments (and nutrients attached with the sediments) and other impurities/debris from the stormwater, stopping those from flowing into the urban drainage systems, waterways and channels. Since a rainwater tank acts as sedimentation basin, it is likely to trap bigger particles and sediments, which will settle at the bottom of the tank. To avoid the sediment laden water, traditionally the outlet tap from the rainwater tank is provided approximately 100~150 mm above the tank bottom and this space (i.e. 100~150 mm) is kept to accommodate the accumulated sediments and debris. However, in reality, if the rainwater water is used for garden irrigation, then such sediment laden water is beneficial for the plants as it contains nutrients which are helpful for the plant growth. Regular maintenance and cleaning of rainwater tank is necessary, otherwise such accumulated sediments and debris may clog the outlet.

References

Imteaz, M.A., Ahsan, A., Rahman, A. and Mekanik, F. (2013). Modelling stormwater treatment systems using MUSIC: Accuracy. Resources, Conservation & Recycling 71: 15–21.

Melbourne Water (2005). WSUD Engineering Procedures: Stormwater. 978-0-643-09223-5, CSIRO Publishing. Melbourne, Australia.

Manning's '*n*' values for Channels (Chow, 1959)

Type of Channel and Description	Minimum	Normal	Maximum
Natural streams – minor streams (top width at floodstage < 100 ft)			
1. Main Channels			
a. Clean, straight, full stage, no rifts or deep pools	0.025	0.030	0.033
b. Same as above, but more stones and weeds	0.030	0.035	0.040
c. Clean, winding, some pools and shoals	0.033	0.040	0.045
d. Same as above, but some weeds and stones	0.035	0.045	0.050
e. Same as above, lower stages, more ineffective slopes and sections	0.040	0.048	0.055
f. Same as "d" with more stones	0.045	0.050	0.060
g. Sluggish reaches, weedy, deep pools	0.050	0.070	0.080
h. Very weedy reaches, deep pools, or floodways with heavy stand of timber and underbrush	0.075	0.100	0.150
2. Mountain Streams, No Vegetation in Channel, Banks usually Steep, Trees and Brush along Banks Submerged at High Stages			
a. Bottom: gravels, cobbles, and few boulders	0.030	0.040	0.050
b. Bottom: cobbles with large boulders	0.040	0.050	0.070
3. Floodplains			
a. Pasture, no brush			
1. Short grass	0.025	0.030	0.035

2. High grass	0.030	0.035	0.050
b. Cultivated areas			
1. No crop	0.020	0.030	0.040
2. Mature row crops	0.025	0.035	0.045
3. Mature field crops	0.030	0.040	0.050
c. Brush			
1. Scattered brush, heavy weeds	0.035	0.050	0.070
2. Light brush and trees, in winter	0.035	0.050	0.060
3. Light brush and trees, in summer	0.040	0.060	0.080
4. Medium to dense brush, in winter	0.045	0.070	0.110
5. Medium to dense brush, in summer	0.070	0.100	0.160
d. Trees			
1. Dense willows, summer, straight	0.110	0.150	0.200
2. Cleared land with tree stumps, no sprouts	0.030	0.040	0.050
3. Same as above, but with heavy growth of sprouts	0.050	0.060	0.080
4. Heavy stand of timber, a few down trees, little undergrowth, flood stage below branches	0.080	0.100	0.120
5. Same as 4. with flood stage reaching branches	0.100	0.120	0.160
4. Excavated or Dredged Channels			
a. Earth, straight, and uniform			
1. Clean, recently completed	0.016	0.018	0.020
2. Clean, after weathering	0.018	0.022	0.025
3. Gravel, uniform section, clean	0.022	0.025	0.030
4. With short grass, few weeds	0.022	0.027	0.033
b. Earth winding and sluggish			
1. No vegetation	0.023	0.025	0.030
2. Grass, some weeds	0.025	0.030	0.033
3. Dense weeds or aquatic plants in deep channels	0.030	0.035	0.040
4. Earth bottom and rubble sides	0.028	0.030	0.035

5. Stony bottom and weedy banks	0.025	0.035	0.040
6. Cobble bottom and clean sides	0.030	0.040	0.050
c. Dragline-excavated or dredged			
1. No vegetation	0.025	0.028	0.033
2. Light brush on banks	0.035	0.050	0.060
d. Rock cuts			
1. Smooth and uniform	0.025	0.035	0.040
2. Jagged and irregular	0.035	0.040	0.050
e. Channels not maintained, weeds and brush uncut			
1. Dense weeds, high as flow depth	0.050	0.080	0.120
2. Clean bottom, brush on sides	0.040	0.050	0.080
3. Same as above, highest stage of flow	0.045	0.070	0.110
4. Dense brush, high stage	0.080	0.100	0.140
5. Lined or Constructed Channels			
a. Cement			
1. Neat surface	0.010	0.011	0.013
2. Mortar	0.011	0.013	0.015
b. Wood			
1. Planed, untreated	0.010	0.012	0.014
2. Planed, creosoted	0.011	0.012	0.015
3. Unplaned	0.011	0.013	0.015
4. Plank with battens	0.012	0.015	0.018
5. Lined with roofing paper	0.010	0.014	0.017
c. Concrete			
1. Trowel finish	0.011	0.013	0.015
2. Float finish	0.013	0.015	0.016
3. Finished, with gravel on bottom	0.015	0.017	0.020
4. Unfinished	0.014	0.017	0.020
5. Gunite, good section	0.016	0.019	0.023
6. Gunite, wavy section	0.018	0.022	0.025

7. On good excavated rock	0.017	0.020	
8. On irregular excavated rock	0.022	0.027	
d. Concrete bottom float finish with sides of:			
1. Dressed stone in mortar	0.015	0.017	0.020
2. Random stone in mortar	0.017	0.020	0.024
3. Cement rubble masonry, plastered	0.016	0.020	0.024
4. Cement rubble masonry	0.020	0.025	0.030
5. Dry rubble or riprap	0.020	0.030	0.035
e. Gravel bottom with sides of:			
1. Formed concrete	0.017	0.020	0.025
2. Random stone mortar	0.020	0.023	0.026
3. Dry rubble or riprap	0.023	0.033	0.036
f. Brick			
1. Glazed	0.011	0.013	0.015
2. In cement mortar	0.012	0.015	0.018
g. Masonry			
1. Cemented rubble	0.017	0.025	0.030
2. Dry rubble	0.023	0.032	0.035
h. Dressed ashlar/stone paving	0.013	0.015	0.017
i. Asphalt			
1. Smooth	0.013	0.013	
2. Rough	0.016	0.016	
j. Vegetal lining	0.030		0.500

Reference

Chow, V.T. (1959). Open-Channel Hydraulics. McGraw-Hill, New York. ISBN 10: 007085906X, ISBN13: 9780070859067

Discharge-Velocity graph for concrete pipe (Manning's *n* = 0.012)

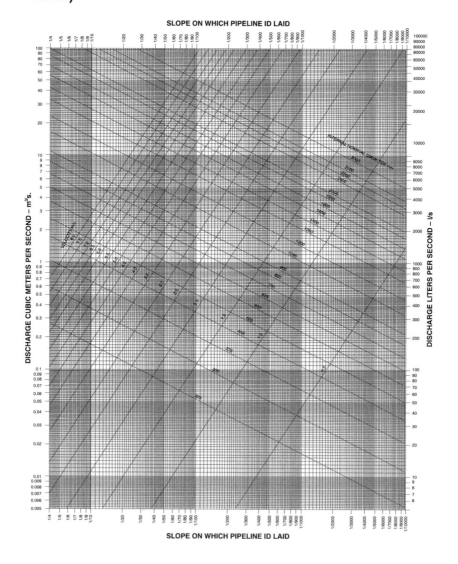

Index

Color Section

Chapter 2: Fig. 2.9, p. 19

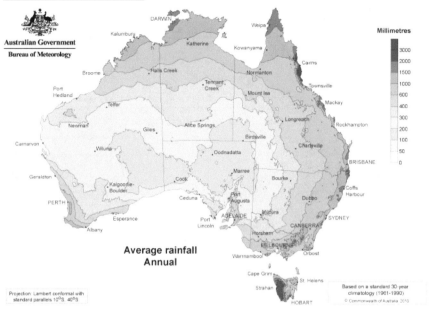

Average rainfall
Annual

Chapter 2: Fig. 2.14, p. 25

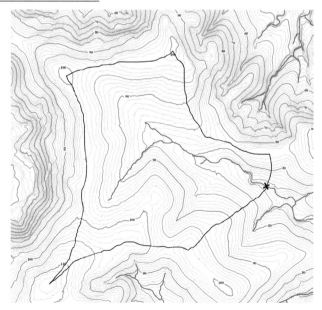

Chapter 4: Fig. 4.5, p. 70

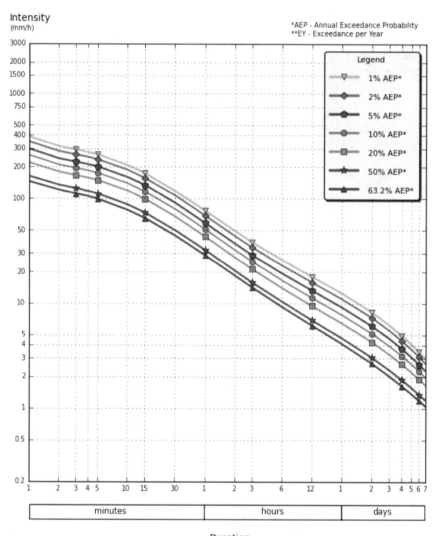

Duration

Chapter 4: Fig. 4.6, p. 71

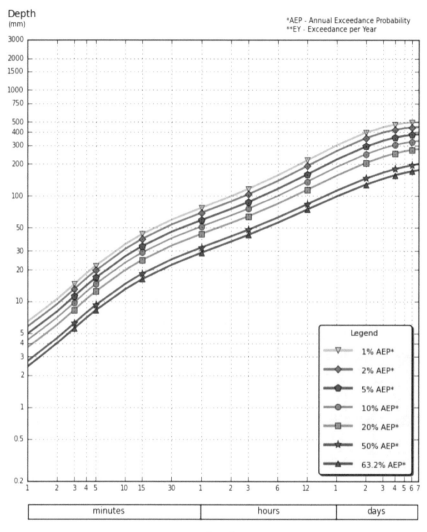

Duration

Chapter 4: Fig. 4.7, p. 73

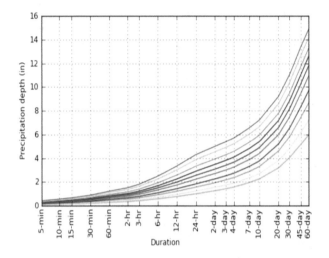

Chapter 4: Fig. 4.8, p. 73

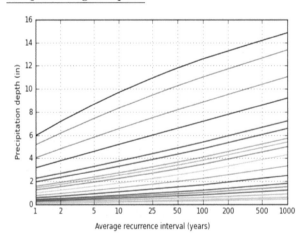

Chapter 11: Fig. 11.21, p. 245

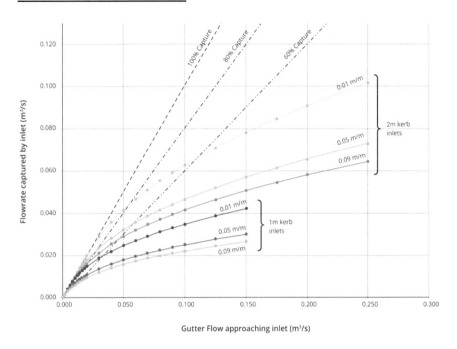

Chapter 12: Fig. 12.2, p. 255